PROTEIN STRUCTURE

N.J. Darby and T.E. Creighton

European Molecular Biology Laboratory, Meyerhofstrasse 1,
6900 Heidelberg, Germany

OXFORD UNIVERSITY PRESS
Oxford New York Tokyo

Oxford University Press, Walton Street, Oxford OX2 6DP

Oxford New York Toronto
Delhi Bombay Calcutta Madras Karachi
Kuala Lumpur Singapore Hong Kong Tokyo
Nairobi Dar es Salaam Cape Town
Melbourne Auckland Madrid
and associated companies in
Berlin Ibadan

Oxford is a trade mark of Oxford University Press

In Focus is a registered trade mark of the Chancellor, Masters, and Scholars of the University of Oxford trading as Oxford University Press

Published in the United States
by Oxford University Press Inc., New York

A catalogue record for this book is available from the British Library

Library of Congress Cataloging in Publication Data
Darby, N.J.
Protein structure / N.J. Darby and T.E. Creighton.
(In focus)
Includes bibliographical references.
1. Proteins–Structure. I. Creighton, Thomas E., 1940–
II. Title. III. Series: In focus (Oxford, England)
QP551.D3 1993 574.19'245–dc20 92–27978
ISBN 0–19–963310–X (pbk.)

Typeset by Footnote Graphics, Warminster, Wiltshire
Printed by Interprint, Malta

Cover illustration courtesy of Dr Georg E. Shulz

PROTEIN STRUCTURE

IN FOCUS

Titles published in the series:

* Antigen-presenting Cells
* Complement
DNA Replication
DNA Topology
Enzyme Kinetics
Gene Structure and Transcription 2nd edition
Genetic Engineering
* Immune Recognition
* B Lymphocytes
* Lymphokines
Membrane Structure and Function
Molecular Basis of Inherited Disease 2nd edition
Molecular Genetic Ecology
Protein Biosynthesis
Protein Degradation
Protein Engineering
Protein Structure
Protein Targeting and Secretion
Regulation of Enzyme Activity
* The Thymus

* Published in association with the British Society for Immunology.

Series editors

David Rickwood

Department of Biology, University of Essex, Wivenhoe Park, Colchester, Essex CO4 3SQ, UK

David Male

Institute of Psychiatry, De Crespigny Park, Denmark Hill, London SE5 8AF, UK

Preface

The last 10 years have seen rapid advances in our understanding of protein structure, largely as a result of the application of protein engineering techniques. These have made it possible to produce large quantities of protein of known amino acid sequence, and to alter that sequence at will to study how the protein functions and how its complex three-dimensional structure is maintained. Concurrently many more protein three-dimensional structures have been determined, due to technical advances in X-ray crystallography and the ability to determine structures by NMR without the need to crystallize the protein. These results have demonstrated that the repertoire of distinct protein folds is smaller than might have been anticipated, giving hope to the quest to understand and predict them. The ability to determine protein primary structures by gene sequencing has resulted in an even greater explosion of predicted protein sequences, and has circumvented the problems involved in the direct sequencing of some proteins, especially rare ones. This has increased the incentive to uncover the structural and functional properties of proteins from their sequences alone. The aim of this volume is to provide a general overview of protein structures and of the factors that stabilize them as well as to examine how protein structures are adapted to carry out their functional roles. General considerations, rather than particular structures, are emphasized, as they are of central importance in the remaining challenges of protein structure research, such as designing proteins *de novo*, predicting protein structure and function from the amino acid sequence alone, and understanding how proteins fold. Nevertheless, the authors hope that readers will also become familiar with some of the more common protein folds described here, and that any mysteries surrounding such terms as 'Greek keys', 'jellyrolls', and 'TIM barrels' will be revealed.

N.J. Darby
T.E. Creighton

Contents

3. Proteins in solution and in membranes

4. Ligand binding and protein function

Abbreviations

Å	ångstrom unit = 10^{-10} m (0.1 nm)
AMP	adenosine 5′-monophosphate
ATP	adenosine 5′-triphosphate
B	crystallographic temperature factor
BPTI	bovine pancreatic trypsin inhibitor
C terminus	carboxy terminus
cDNA	complementary deoxyribonucleic acid
C_α	alpha carbon
C_p	partial heat capacity
ΔC_p	change in heat capacity
e	electrostatic charge
FAD	flavin adenine dinucleotide (oxidized form)
GP	glycogen phosphorylase
ΔG_{bind}	binding free energy
$\Delta\Delta G_{bind}$	change in binding free energy
ΔG_{fold}	free energy of protein folding
hsp	heat-shock protein
h-t-h	helix–turn–helix
IDH	isocitrate dehydrogenase
K_a	association constant
K_d	dissociation constant
$K_{eq,fold}$	equilibrium constant for folding
LDH	lactate dehydrogenase
N terminus	amino terminus
NAD^+	nicotinamide adenine dinucleotide (oxidized form)
$NADP^+$	nicotinamide adenine dinucleotide phosphate (oxidized form)
NOE	nuclear Overhauser effect
PDI	protein disulphide isomerase
pK	negative logarithm of H^+ dissociation constant
PPIase	peptidyl prolyl isomerase
R	the gas constant
R state	relaxed state of an allosteric protein
RNase	ribonuclease
T	absolute temperature

TIM	triose phosphate isomerase
T state	tense state of an allosteric protein

Abbreviations for amino acids

Ala	alanine
Arg	arginine
Asn	asparagine
Asp	aspartic acid
Cys	cysteine
Glu	glutamic acid
Gln	glutamine
Gly	glycine
His	histidine
Ile	isoleucine
Leu	leucine
Lys	lysine
Met	methionine
Phe	phenylalanine
Pro	proline
Ser	serine
Thr	threonine
Trp	tryptophan
Tyr	tyrosine
Val	valine

1

Basic aspects of polypeptide structure

1. The covalent structures of proteins

Proteins are linear polymers of the 20 common L-amino acids whose structures are shown in *Figure 1.1*. Condensation of the α-carboxyl group of one amino acid and the α-amino group of another results in formation of the peptide bond (scheme 1.1).

$$^{+}H_3N-\underset{H}{\overset{R_1}{C}}-CO_2^- + {}^{+}H_3N-\underset{H}{\overset{R_2}{C}}-CO_2^- \longrightarrow {}^{+}H_3N-\underset{H}{\overset{R_1}{C}}-\overset{O}{\overset{\|}{C}}-\underset{H}{N}-\underset{H}{\overset{R_2}{C}}-CO_2^- + H_2O \quad (1.1)$$

The peptide bond is a stable linkage that is substantially hydrolysed only by concentrated acids or bases, at high temperatures, or by proteolytic enzymes (1).

Formation of the peptide bond *in vivo* is a complex enzyme-catalysed event that takes place on the ribosome, in which the genetic information specifies the sequence in which the amino acids are linked (2). Almost all natural proteins contain only the L-isomers of the 20 common amino acids, a restriction imposed by the specificity of the biosynthetic apparatus, although they can be linked together in an enormous number of ways. Further diversity is often introduced post-translationally by a wide variety of covalent modifications of the amino acid side chains (3). In addition, one other amino acid, selenocysteine, can be incorporated into a few proteins directly in response to a codon (UGA) that normally specifies chain termination; sequences on either side of this codon are believed to specify its translation as selenocysteine (4).

Peptides and small proteins can also be synthesized chemically (5). This is less restrictive than biosynthesis in terms of the types of amino acids that can be utilized, and any D- or unusual amino acid may be incorporated into the peptide chain. Current technology enables polypeptide chains of as many as 100–150 amino acids in the length to be prepared in homogeneous form, although considerable effort is required for synthesis of longer chains.

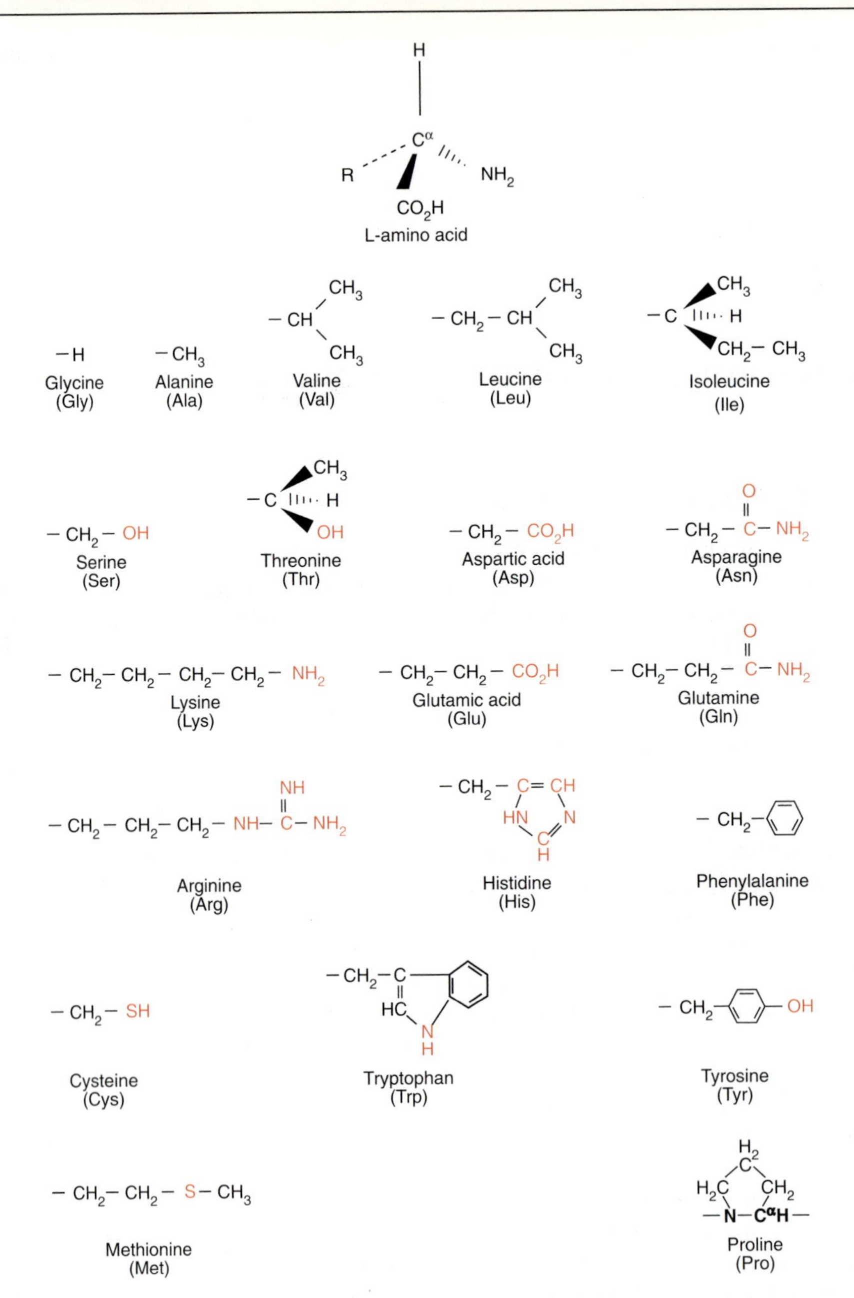

Figure 1.1. The chemical structure of the 20 common amino acids. The general structure of an L-amino acid is shown at the top. The side chain (R) groups of the amino acids are shown below, with the polar functional groups in orange. Note that the representation of proline also shows a part of the peptide backbone (in bold) due to the cyclic structure of this residue.

Once assembled, the polypeptide chain of a protein can be represented as

$$\mathrm{H{-}(NH{-}\overset{\overset{\displaystyle R_i}{|}}{C}H{-}\overset{\overset{\displaystyle O}{\|}}{C})_{\mathit{n}}OH} \tag{1.2}$$

where n is the total number of amino acids linked together and R_i is the side chain of the ith amino acid, which depends upon the sequence in which the amino acids were linked together. The atoms in brackets originate from the same amino acid and are referred to as the amide (—NH—), the alpha carbon ($C_\alpha H$), the carbonyl (—CO—), and the side chain (R). The atoms of the side chain are usually designated as β, γ, δ, ε, ζ, and η in order away from the backbone. Proline residues do not conform with scheme 1.2 in that their side chains are cyclized to the nitrogen atom of the backbone.

2. Non-covalent forces in protein structure

In addition to the covalent structure, non-covalent interactions between different atoms are especially important in defining and stabilizing the three-dimensional structures of large molecules like proteins, in which atoms distant in the covalent structure can interact at close range. These interactions are generally considered to be of four types: *electrostatic interactions* between charges and dipoles, *van der Waals forces*, *hydrogen bonds*, and *hydrophobic interactions*. The strengths of the first three interactions can be measured or calculated in simple situations, such as between two atoms in a vacuum. In the complex environment of a protein in solution or a membrane, however, the strengths of the individual interactions can vary enormously and can be very difficult to assess, due to the multiplicity of atoms present and their varying degrees of conformational freedom. For this reason, it is inappropriate to give standard values for the strengths of the four types of interactions, although average values are frequently quoted.

2.1 Electrostatic interactions

A number of amino acid side chains of proteins, plus the amino (N) and carboxyl (C) termini (scheme 1.1), carry a formal electrostatic charge ($\pm e$) at physiological pH values (*Table 1.1*). In addition, other atoms of some amino acid side chains and of the peptide bond unit itself (see Section 5.1.1) carry partial charges as a result of charge asymmetry due to the different electronegativities of the atoms. This produces a dipole, which is measured as a dipole moment. For example, the side chain oxygen of a serine residue (*Figure 1.1*) can be considered to have a partial charge of $-0.3\,e$, with a corresponding positive charge on the hydrogen atom bonded to it. The magnitude of the well-known Coulombic interaction between two point charges is inversely proportional to the distance between them, but no such simple relationship applies to the interactions involving

Table 1.1. Properties of amino acid residues

Amino acid residue	van der Waals volume (Å^3)	Accessible surface area (Å^2)[a]		Intrinsic pK of ionizable group	Hydrophilicity of side chain (kJ/mol)[b]	Hydrophobicity (kJ/mol)	
		Total residue	Side chain polar atoms			Side chain analogues[b]	*N*-acetyl amides[c]
Alanine	67	113			−1.89	−3.65	−1.30
Arginine	148	241	107	12.0	−93.66	66.61	4.24
Asparagine	96	158	69		−50.69	21.83	2.52
Aspartic acid	91	151	58	3.9–4.0	−56.02	40.57	3.23
Cysteine	86	140	69	9.0–9.5	−15.24	−1.42	−6.46
Glutamine	114	189	91		−49.43	27.21	0.92
Glutamic acid	109	183	77	4.3–4.5	−53.05	32.55	2.69
Glycine	48	85			0	0	0
Histidine	118	194	49	6.0–7.0	−53.17	23.52	−0.54
Isoleucine	124	182			−1.00	−16.71	−7.56
Leucine	124	180			−0.46	−16.71	−7.14
Lysine	135	211	48	10.4–11.1	−50.02	27.25	4.15
Methionine	124	204	43		−16.25	−5.92	−5.17
Phenylalanine	135	218			−13.23	−8.57	−7.52
Proline	90	143					−3.02
Serine	73	122	36		−31.29	18.23	0.17
Threonine	93	146	28		−30.53	14.74	−1.09
Tryptophan	163	259	27		−34.73	−5.84	−9.45
Tyrosine	141	229	43	10.0–10.3	−35.7	4.54	−4.03
Valine	105	160			−1.68	−13.02	−5.12

[a] Measured in a Gly-X-Gly tripeptide in an extended conformation [Miller, S., Janin, J., Lesk, A.M., and Chothia, C. (1987) *J. Mol. Biol.*, **196**, 641].

[b] Hydrophilicity was measured by the partition coefficient of the model for each side chain (backbone replaced by hydrogen atom) from vapour → water, hydrophobicity from water → cyclohexane. For ionizing side chains, the values were corrected for the fraction of each side chain that is ionized at pH 7. Both scales were normalized to zero for the value for Gly [Radzicka, A. and Wolfenden, R. (1988) *Biochemistry*, **27**, 1664].

[c] Measured from the partition coefficient between water and octanol of the *N*-acetyl amino acid amides [Fauchère, J. and Pliska, V. (1983) *Eur. J. Med. Chem.*, **18**, 369].

dipoles, because each dipole consists of at least two effective charges. The strengths of all electrostatic interactions in a homogeneous environment are dependent upon the value of the dielectric constant of the environment, but there is no single appropriate value for the dielectric constant of the heterogeneous environment of a protein molecule (6).

2.2 *van der Waals interactions*

These non-specific interactions occur between all atoms. They result from a complex quantum mechanical phenomenon that can be described, in a very simplified manner, as the transient asymmetry of the electron distribution around one atom inducing a complementary asymmetry in an adjacent atom. The resulting attractive force between them draws the two atoms together. The magnitude of the attraction is critically dependent upon the separation of the two atoms and decreases with the sixth power of the distance between their centres. However, as the atoms approach each other the attractive forces are countered by repulsion between the electron clouds of the two atoms, and there is no net attraction at a separation defined as the sum of their *van der Waals radii.*

2.3 *Hydrogen bonds*

Hydrogen bonds result from the partial sharing of a hydrogen atom between two electronegative atoms. The electron of the hydrogen atom is substantially delocalized on to the atom to which it is formally covalently bound (the donor, D), so the hydrogen atom carries a substantial positive charge. If another atom carrying a local negative charge (the acceptor, A) is in the vicinity of this hydrogen atom, a favourable electrostatic interaction occurs between them.

$$\mathrm{D}\text{——}\mathrm{H} + \mathrm{A} \longleftrightarrow \mathrm{D}^{\delta-} \cdots \mathrm{H}^{\delta+} \cdots\cdot \mathrm{A}^{\delta-} \qquad (1.3)$$

The hydrogen atom is special in this regard because of its small size, on which a substantial charge can be localized. In strong hydrogen bonds, partial transfer of electrons gives a covalent aspect to the hydrogen bond (7).

In proteins the predominant hydrogen bonds involve oxygen and nitrogen atoms, both as donors and acceptors. The hydrogen bonds most important in maintaining protein structure are those between the peptide backbone –NH groups and the backbone carbonyl groups.

Hydrogen bonds in proteins generally have nitrogen or oxygen atoms between 2.6 and 3.1 Å apart, depending somewhat upon the groups involved. Generally, the strongest hydrogen bonds are those that have the shortest bond lengths, but there is a continuum of very weak interactions at long distances. The bond strength is also thought to be maximized when the hydrogen atom is aligned in a straight line between the participating donor and acceptor atoms; the tendency for this to occur could reflect the intrinsic properties of the hydrogen bond or could simply result from minimal steric clashes between the atoms (8).

2.4 *Hydrophobic interactions*

Hydrophobic interactions are generally considered to occur between *non-polar* (i.e. apolar or hydrophobic) groups of atoms in water. The hydrophobicities of molecules are usually determined from their free energy of transfer from water to an organic solvent; the hydrophobicity of a group of atoms is measured by the difference in free energy of transfer of two molecules differing only by this group. The values measured depend somewhat upon the non-polar solvent, the molecules, and the methodologies used. There is reasonable agreement between the relative hydrophobicities of the amino acid side chains measured in different ways, although scales measured using analogues of only the side chains are more extreme than those using molecules that also contain the polar atoms of the backbone [*Table 1.1*; (9)]. The free energy of transfer to non-polar environments is favourable for non-polar groups, but unfavourable for polar groups, which interact favourably with water (10). Consequently, non-polar groups tend to interact with each other, rather than with water, which is the basis of the hydrophobic interaction.

The hydrophobic interaction has unusual properties, which have made it difficult to understand. For example, it is very temperature-dependent and, in the physiological range, increases in magnitude with increasing temperature, whereas other types of interactions decrease in strength. This unusual temperature dependence is a consequence of the way in which water molecules solvate non-polar groups. Understanding the nature of this solvation is central to understanding the hydrophobic effect (11,12).

Aqueous solutions of non-polar molecules are distinguished by having very high heat capacities, which are a measure of the temperature dependence of their enthalpy and entropy. This is thought to be due to the presence around non-polar surfaces in water of hydrogen-bonded ordered arrays of water molecules, also known as *clathrates* or *icebergs*, that are in equilibrium with the more usual liquid state water. The ordered arrays of water have the more negative enthalpy, compensated for by a more positive entropy, and are increasingly populated at lower temperatures.

It was originally thought that the presence of these ordered arrays of water molecules was responsible for the low solubilities of non-polar molecules in water, in that they ‘squeeze out’ the non-polar atoms. More recently it has been realized that the ordered arrays actually improve the interaction of water with non-polar surfaces, increasing the aqueous solubilities of non-polar molecules. Water ordering is more prevalent at low temperatures, so non-polar molecules are more readily solvated by water at low temperatures. This is the reason why the hydrophobic interaction decreases in magnitude at low temperature (11,12).

The term hydrophobic might be taken to imply that there is a repulsive, unfavourable interaction between non-polar surfaces and water. It is clear, however, from the free energy of transfer from the gas phase to aqueous solution that such interactions are energetically favourable, especially at lower temperature. On the other hand, they are less favourable than those between non-polar molecules themselves and between water molecules. The magnitude of the hydrophobic interaction is the net difference between the van der Waals interactions between non-polar molecules and the less favourable, but temperature-sensitive, interaction between non-polar molecules and water.

3. The amino acid residues

The properties of the amino acid side chains (*Figure 1.1*) have a major influence on how protein structures arise and are stabilized. Some of the most important properties of the 20 common amino acid residues are summarized in *Table 1.1*. Key features to recognize are the charge, hydrogen bonding potential, overall size, and relative hydrophobicity of the side chain.

Glycine has no side chain, which gives the polypeptide chain special flexibility (see Section 5.1.2). Conversely, the cyclic side chain of proline limits the flexibility of the polypeptide backbone at this residue.

In general terms, the side chains of alanine, isoleucine, leucine, methionine, phenylalanine, and valine residues are considered hydrophobic. The most important properties of these residues are their shapes, volumes, and flexibilities. Tryptophan and tyrosine residues are often regarded as hydrophobic, but the nitrogen atom of the indole ring of tryptophan and the hydroxyl group of tyrosine give them some polar properties. The aromatic side chains of phenylalanine, tryptophan, and tyrosine residues have recently been recognized as having significant potential for electrostatic interactions, because their π electrons are concentrated on the faces of the aromatic rings, with an electron deficit on the ring hydrogen atoms (13).

Asparagine, glutamine, serine, and threonine side chains possess hydroxyl or amide groups capable of forming hydrogen bonds, and these residues are considered polar. Lysine and arginine side chains have intrinsic pK values of 10.4–11, and 12, respectively, so they carry a net positive charge at physiological pH. Aspartic and glutamic acids are negatively charged because their side chain carboxyl groups have intrinsic pKs of 4–4.5. Note, however, that although the terminal groups of all of these side chains are polar, they are always separated from the C_α atom and the backbone by a number of hydrophobic $-CH_2$ units (see Chapter 2, Section 3.2).

Cysteine residues are important heavy metal binding sites in a number of proteins. The thiol group of cysteine has a pK of between 9 and 9.5 and the thiolate anion, although not normally present to a great extent at physiological pH, is very reactive. Pairs of Cys residues also have the potential to form disulphide bonds in which case the product is a cystine residue. Disulphide bonds can be formed by air oxidation:

$$2(R\text{—}CH_2SH) + \tfrac{1}{2}O_2 \longrightarrow R\text{—}CH_2S\text{—}SCH_2\text{—}R + H_2O \qquad (1.4)$$

or thiol–disulphide interchange with a disulphide reagent like oxidized glutathione (GSSG):

$$2(R\text{—}CH_2SH) + GSSG \longrightarrow R\text{—}CH_2S\text{—}SCH_2\text{—}R + 2GSH \qquad (1.5)$$

Disulphide bonds are covalent, but reversible, cross-links that are often important in maintaining the structures of proteins (see Chapter 2, Section 3.5).

Histidine residues are important in the active sites of a number of enzymes, as the pK of the imidazole side chain is in the physiological range, allowing it to exist in both the ionized and non-ionized states. In the non-ionized state, one of the nitrogen atoms can exist as a donor for hydrogen bonds, the other as an acceptor. With two similar nitrogen atoms on its side chain, and a hydrogen atom almost always bonded to one of them, histidine residues are ideally suited to participate in proton transfers at the active sites of enzymes. Histidine residues can also bind metals and are involved in the binding of a number of metal-containing prosthetic groups, such as haem.

4. Primary structures of proteins

Proteins exhibit a wide range of sizes, some having peptide chains up to several thousand amino acid residues long. Indeed, the muscle protein titin has a mass of about 3×10^6 daltons, corresponding to approximately 27 000 amino acids. This is exceptional, however, and an average polypeptide chain is about 300–500 residues. The lengths of polypeptide chains are most often determined by SDS gel electrophoresis. This technique relies on the proteins being denatured and uniformly coated with SDS, so that the effects of intrinsic protein charge and shape are minimized (14). In such circumstances, the rate of migration of the SDS-polypeptide complex through a sieving polyacrylamide gel is inversely proportional to the logarithm of the length of the polypeptide. The mass of an unknown polypeptide chain can be estimated by comparing its mobility to those of a series of standard proteins. The mass measured in this way is generally accurate only to within $\pm 10\%$ and can be subject to greater error if the polypeptide chain does not bind the normal amount of SDS because of its amino acid composition or if it has been modified by covalent cross-links or by introduction of bulky groups. Developments in mass spectrometry now permit the mass of a protein to be determined with high accuracy using very small samples (15). In particular, the electrospray ionization technique can determine masses of molecular weight up to 100 000 daltons with an accuracy of about 0.01%, using only nanomole amounts of sample.

The sequence of amino acids in a polypeptide chain is known as the primary structure. It is generally determined directly by sequential degradation of the polypeptide chain (or subfragments of it) from the N terminus by the Edman procedure (16,17), or from the gene or cDNA sequence that encodes the protein. While the gene sequence is usually the easiest and most accurate way of determining the sequence of a protein, analysis of any post-translational modifications must be carried out directly on the protein. The accuracy of mass measurements by mass spectrometry makes it very useful for detecting post-translational modifications of proteins from discrepancies between their measured molecular weights and those predicted from their gene sequences (15).

An enormous number of protein sequences are possible, 20^n, where n is the number of residues. Consequently two long sequences are unlikely to be similar simply by chance. Nevertheless, many natural protein sequences are found to be similar to varying extents; some are virtually identical, while others show barely significant resemblance. Any statistically significant sequence similarity is taken as indicating that the two proteins have a common evolutionary ancestor, in which case they are said to be homologous. The sequence similarities between homologous proteins are very powerful tools in reconstructing the process of evolution (18,19).

Sequence homology is an 'all-or-none' phenomenon; proteins are either homologous or they are not (20). The term 'percentage homology' frequently encountered in the literature usually means 'percentage identity'. Only with mosaic proteins, composed of several segments of polypeptide chain with different origins (21), can it be correct to ascribe fractional homology. Homology is one of the most powerful methods for identifying the functions of newly sequenced proteins or genes (for examples, see Chapter 4), so considerable effort is being expended to develop more rapid and more powerful sequence comparison methods.

The overall amino acid compositions of most proteins are quite similar, in that the 20 amino acids occur at certain frequencies that are similar in most proteins. On the other hand, membrane proteins can often be distinguished because they contain a somewhat higher proportion of hydrophobic amino acids than do soluble proteins. The most markedly abnormal proteins in terms of composition are the fibrous structural proteins, which, because of their regular conformations, have amino acid sequences that tend to be repetitive (see Section 6).

Methionine is found more frequently at the N terminus of proteins than any other residue, as might be expected from it being the initiator amino acid in the biosynthesis of peptide chains. Otherwise, on average, the 20 amino acids are randomly distributed through the polypeptide backbone (22), suggesting there are no simple rules relating protein structure to amino acid sequence.

5. Three-dimensional aspects of polypeptide structure

5.1 Conformations of polypeptides

The covalent structure of a macromolecule is not sufficient to determine its three-dimensional structure, due to the possibility of different rotations about the many covalent bonds. Three-dimensional structures that differ only in this way are referred to as *conformations*. Different conformations should be interconverted solely by bond rotations, although polypeptides differing covalently in their disulphide bonds between cysteine residues can be considered as being different conformations, when conformation dictates the disulphide bond pairings. The basis of the three-dimensional conformations of polypeptide chains is central to understanding the structures and functions of proteins.

5.1.1. The peptide bond

The peptide bond is a relatively rigid unit (*Figure 1.2*) as a consequence of its 40% double bond character, which results from delocalization of the lone pair of electrons on the nitrogen to form a resonance hybrid. This also creates a permanent dipole (23,24):

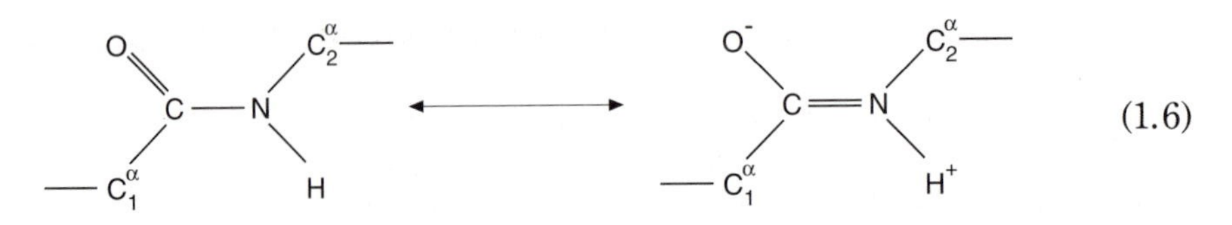

(1.6)

The peptide bond is constrained by its partial double bond nature and so the CO, NH, and both adjacent C_α atoms lie in the same plane. Two planar forms are possible, in which adjacent C_α carbons are fixed in either the *cis* or the *trans* configuration.

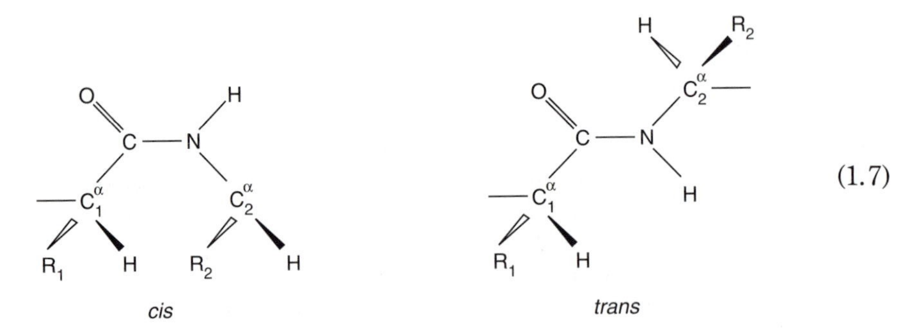

(1.7)

The *trans* form of the bond is intrinsically favoured 1000-fold over the *cis* form because *cis* peptide bonds lead to considerable steric hindrance between adjacent side chains in most cases (25). When the next residue is proline, however, there is little difference between the *cis* and *trans* isomers and they have comparable free energies. In folded proteins about 6% of proline residues are preceded by a *cis* peptide bond.

5.1.2 Polypeptide conformations

The conformation of the polypeptide backbone is defined by the *torsion angles* ϕ, ψ, and ω of each residue (*Figure 1.2*), while those of the side chain are designated by χ_i, where i is the number of the bond starting from the C_α atom. The range of possible values of each is taken to be between $+180°$ and $-180°$ (26). The value of $+180°$ (which is the same as $-180°$) is given to each of the torsion angles in the maximally extended chain. Rotation about the peptide bond is defined by the bond angle ω, which is 0° in the *cis* configuration and $\pm 180°$ in the *trans* configuration.

Only certain values of ϕ and ψ are permitted, because many rotations about these bonds lead to unfavourable steric interactions between atoms of the same and neighbouring residues. The combinations of ϕ and ψ torsion angles in a residue are usually described in *Ramachandran plots*, with the permitted values usually indicated (*Figure 1.3*).

The permitted values of the ϕ and ψ torsion angles depend somewhat upon the

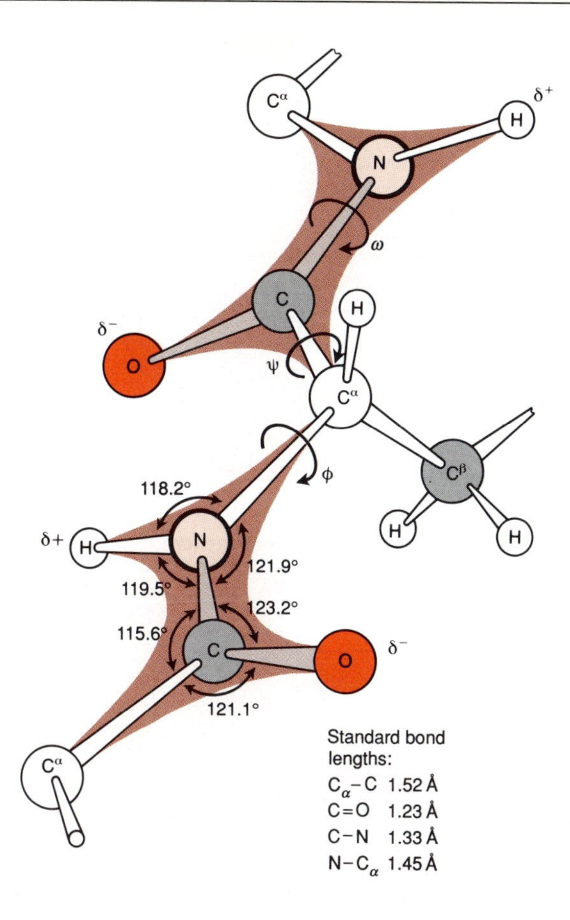

Figure 1.2. The geometry of the peptide backbone with *trans* peptide bonds. The average dimensions and bond angles are determined from crystallographic data. The main chain bond angles, ϕ, ψ, and ω have increasingly positive values in the direction of rotation shown. The shading indicates the planarity of the peptide bond and partial double bonds are shown in colour. The partial charges δ^+ and δ^- show the peptide bond dipole. Adapted from Moody, P.C.E. and Wilkinson, A.J. *In Focus: Protein Engineering* (IRL Press 1990).

size and shape of the side chain. The range of permitted bond angles ϕ and ψ is much the greatest for the glycine residue because of the absence of a side chain. A glycine residue provides flexibility in the polypeptide chain and allows it to make tight turns. Other residues have much less conformational flexibility, and

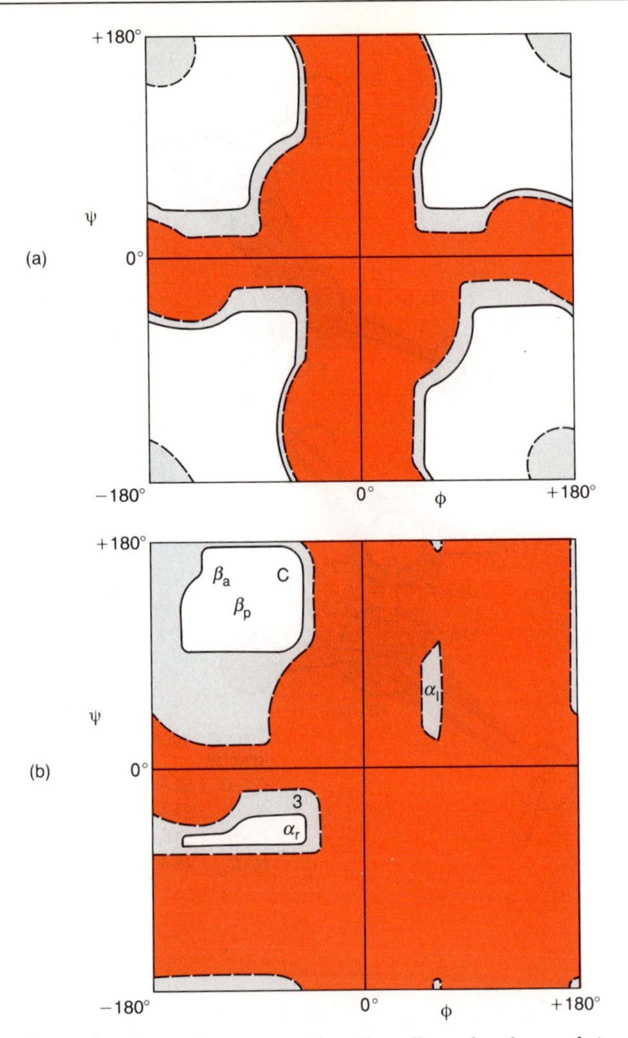

Figure 1.3. Ramachandran plots comparing the allowed values of ϕ and ψ for (**a**) glycine and (**b**) alanine residues. The fully allowed regions are shown in white. The orange coloured areas show the regions disallowed by steric clashes if the atoms are treated as hard spheres. Small decreases in minimum contact distances expand the allowed region to that of the regions shaded in grey. The values observed in proteins generally fall within the expanded allowed regions, which include both the white and the grey shaded areas. Also shown are the approximate values of ϕ and ψ for regular structural features (Section 5.2): α_r and α_l, right- and left-handed α-helices, respectively: β_p and β_a, parallel and anti-parallel β-sheets, respectively: 3, the 3_{10} helix: C, the collagen helix (Section 6.2). Adapted from Moody, P.C.E. and Wilkinson, A.J. *In Focus: Protein Engineering* (IRL Press 1990).

bulky residues, such as valine and isoleucine, can only adopt a more limited range of ϕ and ψ bond angles. Proline residues with their cyclic side chains are only allowed a value for ϕ of −60° ± 20°, depending upon the extent to which the ring system can be distorted.

In the absence of net interactions between atoms distant in the covalent structure and with the solvent, the polypeptide chain would behave as a *random coil* capable of adopting a multitude of conformational states of similar energies. The number of such states is limited by the restriction that two atoms cannot exist in the same space at the same time, but even so, a vast number of conformations remains available. For example, a 100 residue peptide chain in which each residue is allowed to adopt only two conformations (a very conservative estimate) would have over 10^{30} theoretical conformational states. The ability to adopt so many conformations, the *conformational entropy*, is a contributing factor to keeping the peptide chain unfolded. Interactions between groups in the peptide chain and with the solvent, however, will make some conformations have less energy than others. In the case of biological proteins, essentially a single conformation predominates, which requires substantial stabilizing interactions.

5.2 Regular polypeptide structures

In regular polypeptide structures, the same local structure is present sequentially along the polypeptide chain, so there is an overall helical conformation. Each type of regular structure is characterized by the values of the ϕ and ψ torsion angles, by the number of residues per turn of the helix, and by the distance along the helix axis of adjacent residues (*Table 1.2*). This is the next level of protein structure after the primary structure, and is known as *secondary structure.*

5.2.1 The α-helix

The *α-helix* is the most common regular conformation in proteins. It has a characteristic pattern of hydrogen bonds between the backbone carbonyl oxygen atom of each residue and the backbone NH group of the fourth residue along the chain (*Figure 1.4*). This leaves the first three NH groups and the last three carbonyl oxygens of the helix without hydrogen bonds. Although the details are not certain, the α-helix is probably stabilized by the hydrogen bonds, by van der Waals interactions between atoms of the backbone packed together, and by the favourable ϕ and ψ angles. All the hydrogen bonds point in the same direction, so the dipoles of the individual peptide bonds (see Section 5.1.1) interact favourably.

Table 1.2. Parameters for regular structural elements

	Bond angle (degrees)			Residues/turn	Translation/residue (Å)
	ϕ	ψ	ω		
Right-handed α-helix	−57	−47	180	3.6	1.5
Parallel β-sheet	−119	+113	180	2.0	3.2
Anti-parallel β-sheet	−139	+135	−178	2.0	3.4
3_{10} helix	−49	−26	180	3.0	2.0

Adapted from Ramachandran, G.N. and Sasisekharan, V. (1968) *Adv. Protein Chem.*, **23**, 283 and IUPAC-IUB Commission on Biochemical Nomenclature (1970) *Biochemistry*, **9**, 3471.

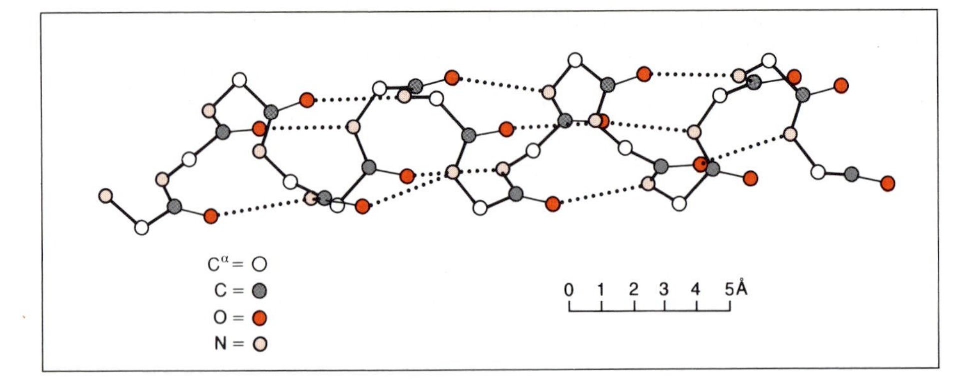

Figure 1.4. An α-helix. The path of the peptide backbone is shown as a bold line, and hydrogen bonds are dotted lines between carbonyl oxygen atoms and NH groups. Reproduced with permission from Moody, P.C.E. and Wilkinson, A.J. *In Focus: Protein Engineering* (IRL Press 1990).

The partial charges left at the ends of the helix are well separated, so the α-helix can be considered to have a substantial dipole moment (23,24).

The geometry of the α-helix arranges the side chains so that they are staggered by 100° looking along the axis of the helix, minimizing possible steric clashes between them. The disposition of side chains on an α-helix is frequently displayed using 'helical wheels', which are projections down the helix axis of the side chains. *Amphipathic helices* have non-polar side chains on one side and polar side chains on the other. The non-polar surfaces of such amphipathic helices tend to interact with each other or with other non-polar surfaces, such as those of membranes. The amphipathic nature of structures is conveniently expressed as the hydrophobic moment, which is analogous to a dipole moment but is proportional to the spatial separation of hydrophobic and polar groups (27).

All amino acid residues are structurally compatible with forming the α-helix, although they differ in their tendencies to do so. The greatest exception is proline, which cannot participate in the normal backbone hydrogen bonding of the helix. Insertion of proline into a helix causes the helix to kink because of the restricted ϕ angle that it can adopt. However, proline is frequently found within the first turn of an α-helix, where its lack of backbone hydrogen bonding can be accommodated and its side chain geometry is quite favourable.

An isolated α-helix is only marginally stable in aqueous solution, because the interactions stabilizing it only partly overcome the conformational entropy of the unfolded state. The α-helical conformation is adopted by only some short peptide sequences, and only at low temperatures (28). The stability of an isolated α-helix depends upon what residues are present, for they vary in their helical propensities. For example, glycine residues diminish the stability of the α-helix because of their flexibility in the unfolded state.

Helix formation tends to be co-operative (see Chapter 3, Section 4.3) and is generally described in homopolypeptides by the *Zimm–Bragg model* (28,29). There is only a small probability (σ), of about 10^{-3}, that four sequential residues will each adopt the conformation required to form the first hydrogen bond and the first turn of the helix. Adding each additional residue to the helix is much easier and is described by the equilibrium constant, s. The overall equilibrium constant for forming a helix of n residues is given by σs^n. The value of s is less than unity for helix-breaking residues, especially proline and glycine, and only somewhat greater than one for even the most favourable residues, such as alanine. Even in favourable cases, the helix is unstable unless the homopolypeptide chain is long, so n has a large value. Although helix formation is very rapid, occurring on the microsecond time scale, the rate of unravelling of an isolated helix is equally fast.

It was once thought that each of the amino acids had a unique value of σ and s, which could be measured by determining the effect on helicity of incorporating a 'guest' amino acid into a 'host' polypeptide. It is now clear, however, that the value of s for any amino acid residue depends at least to some extent on the context in which it occurs in the overall sequence (28).

Using short peptides, it is possible to assess the propensity of the individual amino acids to form α-helices and to examine some of the factors that stabilize the α-helical conformation. Alanine is found to have a particularly high helical propensity; residues with longer side chains are more restricted in their side chain conformations in the helix and are thus less likely to be helical. Negatively and positively charged residues at the N and the C termini of the α-helix, respectively, interact favourably with the helix dipole and thereby stabilize the helix; the opposite locations destabilize the helix. Interactions between side chains can also affect the stability of the α-helix, as when oppositely charged residues are located at intervals of three to four residues, so that they occur on the same face of the helix and can form salt bridges (28).

A left-handed α-helical backbone is also possible, with values of ϕ and ψ the same as those for the right-handed helix, but of opposite sign. Such a conformation is generally not stable, however, as the side chains come into too close contact with the backbone.

5.2.2 Other types of helix

Variations on the α-helix, with hydrogen bonds between residues nearer and further apart, are not stable in isolation. The more tightly coiled helix (*Table 1.2*) is known as the 3_{10} helix, because it has three residues per turn and 10 atoms in the ring defined by the hydrogen bond. This helix is too closely packed, however, and is usually only observed as one turn at the ends of α-helices. The less tightly coiled helices, such as the π-helix with hydrogen bonds between every fifth residue, do not have stabilizing van der Waals interactions between backbone atoms and are not observed.

Polyproline adopts either of two special helical structures, known as the polyproline I and II helices. Both have three residues per turn, but differ in having *cis* and *trans* peptide bonds, respectively.

5.2.3 β-Sheets

The second major element of regular secondary structure is the β-*strand*, which has the backbone extended. It can be considered as a special type of helix, with two residues per turn and a translation of 3.4 Å per residue. This extended structure is not stable in isolation, and is observed only in β-*sheets*, which are aggregates of multiple β-strands, running in the same (parallel) or opposite (anti-parallel) directions. The strands are cross-linked by hydrogen bonds between the carbonyl and NH groups that point out from the peptide chain at right angles to the side chains (*Figure 1.5*). Alternate C_α atoms lie slightly above and slightly below the plane of the sheet, so the path of the backbone has a pleated appearance (*Figure 1.5c*).

The factors that stabilize the β-strand and β-sheet conformations are not certain, for there is no adequate model system of studying their formation (30). Homopolypeptides that form β-sheets tend to make large intermolecular aggregates, for an intramolecular sheet requires not only β-strands but also turns between them (see Chapter 2, Section 3.4). Proline residues do not fit well into β-sheets, as they cannot participate in the hydrogen bonding network. The side chains of neighbouring residues in adjacent strands of a sheet project in the same direction and pack closely together, so a variety of interactions occur there. Like the α-helix, β-sheets can have faces that are predominantly polar or non-polar, by having residues of the same type at alternate positions in each strand.

6. The structures of fibrous proteins

The fibrous proteins are a group of structural proteins that have as their main characteristic an extended regular three-dimensional structure that is specified by a relatively regular amino acid sequence. Their structures are intermediate in complexity between those of isolated regular structural elements (α-helices and β-sheets) and globular proteins (Chapter 2), so they are a good starting point for examining how protein structures are built up and stabilized.

6.1 Silk fibroin

Fibroin is a protein of molecular weight 350 000–415 000 that consists primarily of segments of β-sheet interspersed by 100–200 residue segments with irregular conformations. The amino acid sequence contains about 50 repeated sequences of

-(Gly-Ala)$_2$-Gly-Ser-Gly-Ala-Ala-Gly-(Ser-Gly-Ala-Gly-Ala-Gly)$_8$-Tyr-

sections that are believed to form anti-parallel β-sheets. The glycine residues are believed to point out from one side of the sheet and the serine and alanine residues on the other side. The sheets are thought to be stacked on top of each other, with the glycine faces packing against each other and alternating with faces of alanine and serine in contact. In the direction of the peptide backbone, the structure is strong and inextensible, because deformation would require the breakage of backbone covalent bonds. The sheets are only held together by

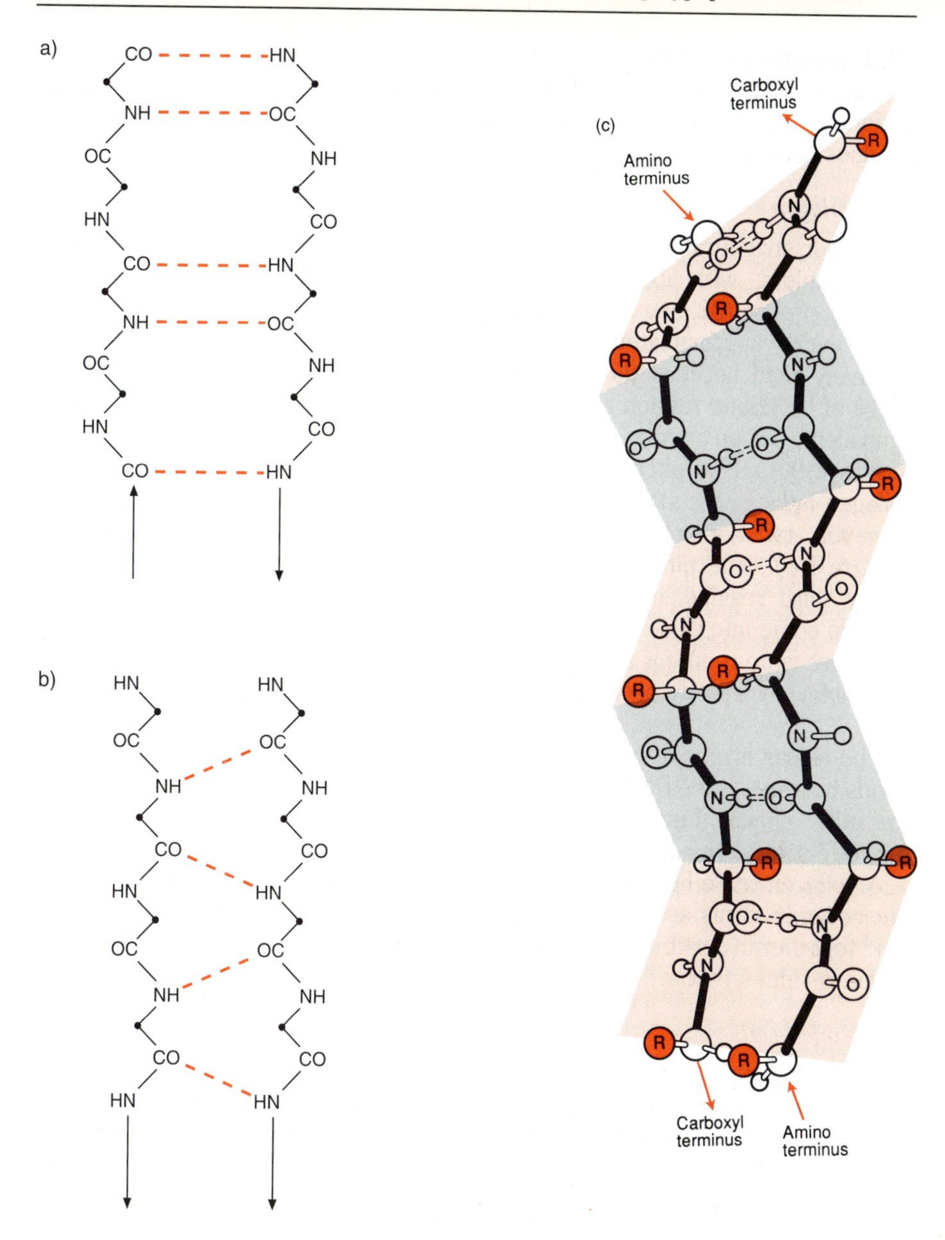

Figure 1.5. Hydrogen bonding patterns in anti-parallel (**a**) and parallel (**b**) β-sheets. The solid black circles are C_α carbon atoms and the dotted lines are hydrogen bonds between backbone carbonyl and NH groups. (**c**) Two strands of an anti-parallel β-sheet with the planes of the bonds highlighted to emphasize the pleated appearance. Note that the side chains of neighbouring residues, denoted as R, project in the same directions. Reproduced with permission from Arnstein, H.R.V. and Cox, R.A. *In Focus: Protein Biosynthesis* (IRL Press 1992).

weak non-specific interactions, however, and can be distorted relatively easily, so silk is flexible. Inclusion of bulky residues breaks up the regular β-sheet structure, and the sheets no longer pack so closely, so the presence of these residues accounts for much of the localized irregular structure in silk (31).

6.2 Collagen

The major structural protein collagen has a repeating primary sequence of (Gly-X-Y)$_n$ triplets, with about a third of the X and Y residues being proline or hydroxy-proline (32,33). The peptide backbone forms a left-handed helix, with ϕ and ψ being approximately $-60°$ and $+140°$, respectively. This structure is relatively rigid because the proline residues can only adopt a very restricted range of backbone torsion angles. Each turn of the collagen helix is not exactly equivalent, but on average there are 3.3 residues per turn, and the pitch of each turn is 2.9 Å. This pitch is nearly twice that of an α-helix, and an individual collagen helix is not stable in isolation. Three of the left-handed helical chains, however, twist about each other to form a stable right-handed *super-helix* (*Figure 1.6*). The pitch of the super-helix is 86 Å, with 10 triplets per turn. Every third residue in each chain must be glycine because residues at this position come into close contact with the other chains along the axis of the helix, and there is no room for a side chain. The side chains of the X and Y residues in the triplet are readily accommodated, because they point away from the helix axis.

The chains are held together by van der Waals interactions and by hydrogen bonds between the NH of the glycine residues and a backbone carbonyl in one of the other chains. Further stability results from hydrogen bonds involving the hydroxy-proline residues.

A collagen triple helix typically has a length of 1000 Å and a diameter of 14 Å, but collagen fibrils are built up of parallel super-helical aggregates. These are held together in part by covalent cross-links involving the side chains of hydroxy-lysine residues that are introduced after the fibrils are formed (34).

6.3 Coiled-coils

A number of structural proteins, such as the intermediate filaments, α-keratin, myosin, fibrinogen, and the so-called 'leucine zippers', adopt similar coiled-coil structures (35,36). Coiled-coils have two, or occasionally three, α-helical strands wound around each other to form a left-handed super-helix. The strands may be identical polypeptide chains or they may be different.

The individual helices of the coiled-coil are amphipathic (see Section 5.2.1), and the non-polar faces pack together along the helix axis and stabilize the structure by hydrophobic interactions. Slight distortions of the individual helices and the inclination of their axes with respect to each other allows the side chains of the apolar residues to mesh together (*Figure 1.7*). Such meshing was originally termed *'knobs into holes'* packing. The left-handed supercoiling of the individual helices about each other allows this packing to be maintained over long distances; an example is the 80 helical turns of tropomyosin.

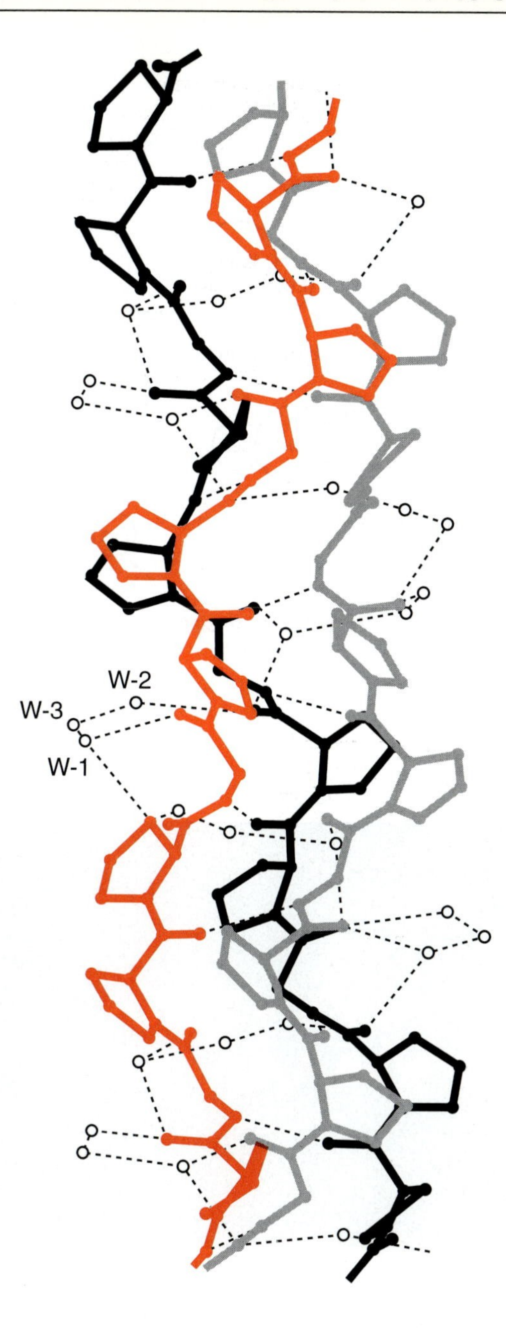

Figure 1.6. The structure of (Pro-Pro-Gly)$_{10}$, which resembles the collagen helix. Three fixed water molecules per repeating Pro-Pro-Gly triplet are indicated by the open circles. The broken lines indicate hydrogen bonds. Reproduced with permission from Okuyama, K., Okuyama, K., Arnott, S., Takayanagi, M., and Kakudo, M. (1981) *J. Mol. Biol.*, **152**, 427.

(a)

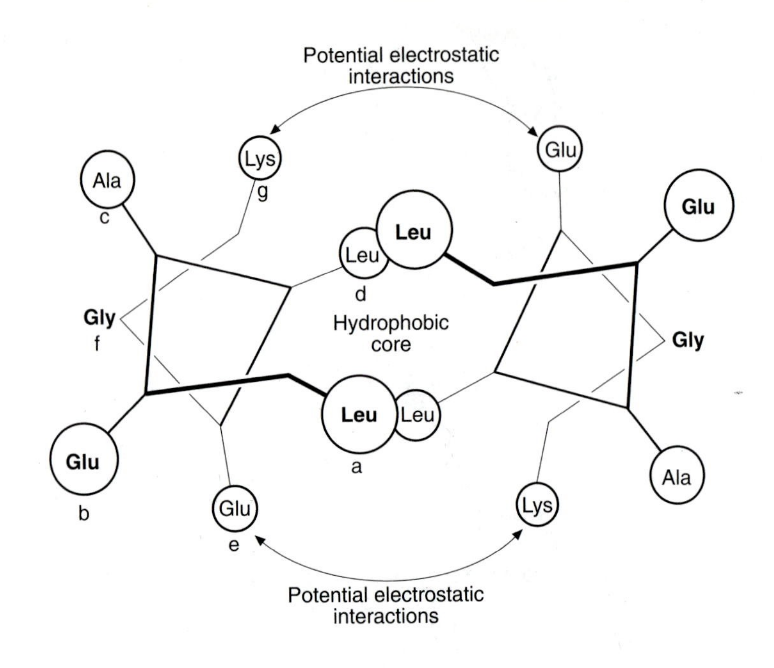

(b)

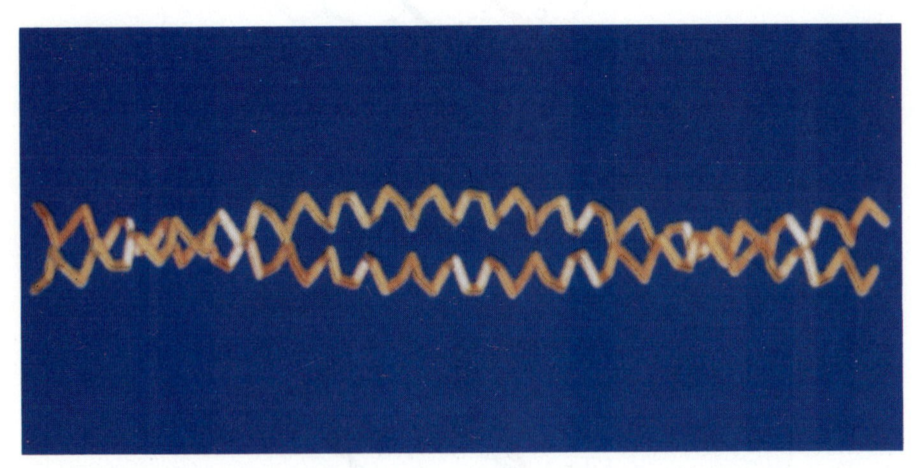

Figure 1.7. The two-chain coiled-coil. (**a**) Idealized view of packing interactions between side chains within one heptad repeat. Reproduced with permission from DeGrado, W.F., Wassermann, Z.R., and Lear, J.D. (1989) *Science*, **243**, 625. (**b**) Path of the polypeptide backbone in a coiled-coil. Reproduced with permission from ref. 36.

The amphipathic nature of the individual helix is readily apparent from the amino acid sequences of these proteins, which exhibit a heptad repeat (*a-b-c-d-e-f-g*). Hydrophobic residues occur at positions *a* and *d*; the former is often leucine, isoleucine, or alanine, whilst the latter is generally leucine or alanine. These residues from different strands pack together in the coiled-coil. On the polar face of the helix, interchain electrostatic interactions often occur between acidic and basic residues at positions *e* and *g*.

7. Further reading

Creighton, T.E. (1993) *Proteins: Structures and Molecular Properties.* W.H. Freeman, New York. Second edition.
Creighton, T.E. (ed.) (1989) *Protein Structure: a Practical Approach.* IRL Press, Oxford.
DeGrado, W.F. (1988) Design of peptides and proteins. *Adv. Protein Chem.*, **39**, 51.
Dickerson, R.E. and Geis, I. (1969) *The Structure and Action of Proteins.* Harper and Row, New York.
Doolittle, R.F. (1986) *Of URFs and Orfs.* University Science Books, Mill Valley, CA.
Israelachvili, J.N. (1985) *Intermolecular and Surface Forces: with Applications to Colloidal and Biological Systems.* Academic Press, New York.
Means, G.E. and Feeney, R.E. (1971) *Chemical Modifications of Proteins.* Holden-Day, San Francisco.

8. References

1. White, R.H. (1984) *Nature*, **310**, 430.
2. Arnstein, H.R.V. and Cox, R.A. (1990) *In Focus: Protein Biosynthesis.* Oxford University Press, Oxford.
3. Wold, F. (1981) *Annu. Rev. Biochem*, **50**, 783.
4. Böck, A., Forchhammer, K., Heider, J., and Baron, C. (1991) *Trends Biochem. Sci.*, **16**, 463.
5. Kent, S.B.H. (1988) *Annu. Rev. Biochem.*, **57**, 957.
6. Sharp, K.A. and Honig, B. (1990) *Annu. Rev. Biophys. Biophys. Chem.*, **19**, 301.
7. Umeyama, H. and Morokuma, K. (1977) *J. Am. Chem. Soc.*, **99**, 1316.
8. Taylor, R. and Kennard, O. (1984) *Acc. Chem. Res.*, **17**, 320.
9. Roseman, M.A. (1988) *J. Mol. Biol.*, **200**, 513.
10. Wolfenden, R. (1983) *Science*, **222**, 1087.
11. Privalov, P.L. and Gill, S.J. (1989) *Pure Appl. Chem.*, **61**, 1097.
12. Privalov, P.L. and Gill, S.J. (1988) *Adv. Protein Chem.*, **39**, 191.
13. Burley, S.K. and Petsko, G.A. (1988) *Adv. Protein Chem.*, **39**, 125.
14. Mattice, W., Riser, J.M., and Clark, D.S. (1976) *Biochemistry*, **15**, 4264.
15. Fenselau, C. (1991) *Annu. Rev. Biophys. Biophys. Chem.*, **20**, 205.
16. Walsh, K.A., Ericsson, L.H., Parmelee, D.C., and Titani, K. (1981) *Annu. Rev. Biochem.*, **50**, 261.
17. Hunkapiller, M.W., Strickler, J.E., and Wilson, K.J. (1974) *Science*, **226**, 304.
18. Wilson, A.C., Carlson, S.S., and White, J.J. (1977) *Annu. Rev. Biochem.*, **46**, 573.
19. Felsenstein, J. (1988) *Annu. Rev. Genet.*, **22**, 521.
20. Reeck, G.R. *et al.* (1987) *Cell*, **50**, 667,
21. Doolittle, R.F. (1985) *Trends Biochem. Sci.*, **10**, 233.

22. Cserzö,M. and Simon,I. (1989) *Int. J. Peptide Protein Res.*, **34**, 184.
23. Wada,A. (1976) *Adv. Biophys.*, **9**, 1.
24. Hol,W.G.H. (1985) *Prog. Biophys. Mol. Biol.*, **45**, 149.
25. Ramachandran,G.N. and Mitra,K. (1976) *J. Mol. Biol.*, **107**, 85.
26. Ramachandran,G.N. and Sasisekharan,V. (1968) *Adv. Protein Chem.*, **23**, 283.
27. Eisenberg,D., Weiss,R.M., and Terwilliger,T.C. (1982) *Nature*, **299**, 371.
28. Scholtz,J.M. and Baldwin,R.L. (1992) *Annu. Rev. Biophys. Biophys. Chem.*, **21**, 95.
29. Zimm,B.H. and Bragg,J.R. (1959) *J. Chem. Phys.*, **31**, 526.
30. Mattice,W.L. (1989) *Annu. Rev. Biophys. Biophys. Chem.*, **18**, 93.
31. Xu,M. and Lewis,R.V. (1990) *Proc. Natl. Acad. Sci. USA*, **87**, 7120.
32. Bornstein,P. and Sage,H. (1980) *Annu. Rev. Biochem.*, **49**, 957.
33. Eyre,D.R. (1980) *Science*, **207**, 1315.
34. Eyre,D.R. (1984) *Annu. Rev. Biochem.*, **53**, 717.
35. Steinert,P.M. and Roop,D.R. (1988) *Annu. Rev. Biochem.*, **57**, 593.
36. Cohen,C. and Parry,D.A.D. (1990) *Proteins: Struct. Funct. Genet.*, **7**, 1.

2

The three-dimensional structures of proteins

1. Introduction

Most natural proteins do not exist as random coils or as the regular repetitive structures of fibrous proteins described in Chapter 1. Instead, each protein is folded into a single specific three-dimensional structure, which is just one of the very many conformations possible, in which segments of α-helices and β-strands (secondary structure) are packed together in various ways. Such structures are dynamic, however, and have localized variations in conformation.

Proteins are usually biologically active only when folded in their native conformations, so understanding their three-dimensional structures is the key to understanding how they function. At the present time, the three-dimensional structure of a protein must be determined experimentally, unless the structure of a closely related protein is known. In the future, however, it should be possible to predict the structure from just the amino acid sequence.

2. Determining protein structure

Due to the complexity of protein structures, their determination requires a considerable amount of experimental data, even though the standard structures and bond geometries of the individual residues are generally assumed. The most detailed structures are obtained from X-ray diffraction data, but NMR spectroscopy can generate structures that are only slightly less detailed. Other simpler techniques, notably CD spectroscopy, provide a more qualitative measure of the average amount of some structural parameters, especially α-helices and β-sheets.

2.1 CD spectroscopy

The polypeptide backbone of L-amino acid residues is chiral and gives an intense CD signal in the far-UV region (180–240 nm) that is dependent on the conformation of the backbone (1). In particular, α-helices, β-sheets, and reverse turns

(Section 3.4) each give characteristically different spectra in this wavelength range. The approximate average content of such structures in a protein can be estimated from its CD spectrum, by comparison to a database of spectra of proteins of known structure. Although often remarkably accurate, such estimates of secondary structure content must be considered tentative, as other aspects of protein structure are now known to contribute to the CD spectrum (2). IR and Raman spectroscopy are also being developed as complementary approaches for the rapid determination of secondary structure content (3).

CD spectra in the wavelength region 240–310 nm reflect primarily the packing of the aromatic side chains and disulphide bonds in the folded three-dimensional structure. Such spectra are sensitive to changes in structure, but it is impossible to infer any detailed structural information from such data. CD spectra are rapidly acquired and are particularly suitable for following changes in protein structure, for example, upon unfolding.

2.2 X-ray diffraction

X-ray crystallography of proteins requires that the protein molecules be organized in a precise crystal lattice, but the preparation of protein crystals of suitable quality is often a difficult task (4). The basis of the X-ray diffraction technique is that electrons of atoms in the crystal scatter an incident X-ray beam to produce a characteristic pattern of reflections that can be observed with an X-ray detector. The intensities of the individual reflections depend upon whether the waves scattered by all the atoms in the crystal recombine in phase to reinforce each other, or out of phase to cancel each other out. X-ray diffraction studies have been aided immensely by the intense X-ray beams of variable wavelength produced by synchrotrons, which has dramatically increased the rate at which diffraction data can be collected (5).

Knowledge of the intensities and phases of individual reflections would permit reconstruction of the original protein molecule in the crystal lattice, but only the intensities can be measured directly. The phases are usually obtained indirectly by *isomorphous replacement*, from the way that one or a few heavy atoms incorporated into the same, isomorphous crystal lattice affect the diffraction pattern in two or more examples. More recent methods use anomalous scattering by sulphur atoms in a protein (or selenium atoms introduced in their place), or by other introduced heavy atoms, to obtain phase information (6).

Once they have been determined, the intensities and phases of all the X-ray reflections are combined in a Fourier transform to produce maps of the electron density. If the protein primary structure is known, these can be interpreted in terms of the arrangements of atoms in the crystal. The *resolution* achieved in such maps is dependent on the amount and quality of the diffraction data included. Resolution of atoms covalently bound requires data to a resolution of at least 1.5 Å, but accurate protein structures are possible with data to 2 Å resolution. At 3 Å resolution, the complete covalent structure of the protein should be apparent as a continuous ribbon of electron density. At lower resolution only dense regular features, such as α-helices and β-sheets, may be apparent. Once approximate phases are known, to produce an initial structure, the phases may be refined using the information in the amplitudes to generate a more accurate electron density map. If the structure of a related protein is known, it can be used as an initial model with which to fit the amplitude data and to generate an initial set of phases that can be refined to determine the new structure (7). This technique, *molecular replacement*, avoids the need to measure phases for the new crystal.

Neutrons can be used in place of X-rays and have the advantage of being scattered by the nucleus, rather than by electrons, so that scattering is less diffuse (8,9). More importantly, hydrogen atoms (^{1}H) have large negative amplitudes and ^{2}H atoms large positive values, so neutron diffraction is able to determine the positions of hydrogen atoms, which are usually not visible in protein electron density maps, and the distinction between ^{1}H and ^{2}H in hydrogen exchange experiments can also be made (Section 4).

The final electron density map is averaged over all the protein molecules of all the various crystals used and over the time required for the diffraction measurements. Disorder of the crystal lattice and mobility in the protein structure tend to smear out the electron density (Section 4).

2.3 NMR spectroscopy

NMR measures the spin properties of ^{1}H atoms in proteins (10), and of ^{13}C and ^{15}N isotopes that can be incorporated into them by biosynthetic labelling. The resonances of individual atoms must be resolved in the spectra, which is the factor limiting the sizes of proteins that can be studied. The current molecular weight limit is about 20 000 daltons. It is necessary to have at least two-dimensional spectra, where the resonances of the various atoms occur on the diagonal, and interactions between pairs of atoms are observed as off-diagonal 'cross-peaks' at positions corresponding to the resonances of the two atoms. Increased resolution is possible with three- and four-dimensional spectra, which are usually obtained from protein labelled with ^{13}C, ^{15}N, or both (11).

The individual resonances must be assigned to specific protons (more correctly, hydrogen atoms) in individual residues. This requires knowledge of the amino acid sequence and relies upon NMR interactions between protons close in the covalent structure, especially between those of the same residues and those adjacent in the sequence. The observed cross-peaks between sequential residues of different types are matched with the known primary structure to identify the specific residues.

Structural information arises primarily from the *nuclear Overhauser effect* (NOE). A pair of protons gives a detectable NOE cross-peak if they are within 5 Å of each other in space, regardless of their separation in the primary structure. The intensity of the NOE cross-peak diminishes with the sixth power of the distance between the protons; although the intensity can also be diminished by other factors, its magnitude puts constraints on this distance. If sufficient data of this kind are available for protons throughout the structure, the data can be used to determine the overall three-dimensional structure (12). The NOE data are generally used to generate a number of independent structures that are consistent with the distance constraints. These structures are displayed superimposed; the relative agreement between them indicates the degree to which the data define the structure. The more NMR data that are included in the structure calculation, the more accurate it is likely to be, but there is no equivalent of the resolution parameter of X-ray diffraction, and the structures of different parts of the molecule may be defined to different extents.

NMR and X-ray diffraction studies of the same protein have generally produced very similar structures. X-ray structures are usually more detailed, but NMR provides dynamic information concerning flexibility and motion within the protein structure, such as the rates of exchange of hydrogens with the solvent and whether aromatic residues are immobilized or free to rotate (Section 4).

3. Structural organization in globular proteins

Protein structure is generally described in terms of five levels. The *primary structure* refers to the sequence of amino acids in the polypeptide chain. *Secondary structure* is regular local conformations, especially α-helices, β-strands, and the turns that link them. *Supersecondary structure* can be used to refer to compact assemblies of elements of secondary structure. *Tertiary structure* describes the overall topology adopted by the polypeptide chain within a domain. *Quaternary structure* refers to the three-dimensional arrangements of the different subunits; it can also be used with independent structural domains in larger proteins.

3.1 Tertiary structure

Most proteins are folded into roughly spherical structures that are highly compact. They do not have perfectly smooth surfaces, however, for their accessible surface areas are about twice that of a sphere of the same size. The torsion angles of most bonds are close to those that are allowed (see Chapter 1, Section 5.1.1), so relatively little conformational strain is incorporated into the three-dimensional structure, unless it is required for functional reasons (13).

To achieve an overall globular shape, the polypeptide chain tends to follow a relatively straight course, often as an α-helix or β-strand, from one side of the globule to the other. These elements of secondary structure are linked together by loops and reverse turns on the surface that allow the peptide chain to undergo abrupt changes in direction. The compactness of the protein results from the close packing of the elements of secondary structure.

3.2 Protein interiors and exteriors

The interior of a protein is defined as containing those atoms that are inaccessible to solvent molecules. The packing of secondary structure elements within the interior of a globular protein primarily involves its non-polar side chains. The polar groups of the backbone in the interior are paired in hydrogen bonds, largely in secondary structure (14). Virtually all the other polar groups, including those of the side chains, are on the surface and exposed to the solvent. The surfaces of soluble proteins are made up of the polar faces of amphipathic α-helices and β-sheets (see Chapter 1, Sections 5.2.1 and 5.2.3), along with the turns and loops.

Packing is surprisingly dense within the interiors of native proteins, with about 75% of the available space filled by atoms (15). This is more dense than the packing of small molecules in a liquid, but is comparable to that in a crystalline structure. Part of the apparently high density of packing arises because so many atoms are joined by covalent bonds, and not all residues within protein interiors are in contact at their van der Waals distances. Alanine, glycine, isoleucine, leucine, phenylalanine, and valine comprise about 63% of the residues in the protein interior, while more polar side chains are found there only rarely. However, it is generally misleading to classify residues as being either exposed or internal because of the diverse composition of their side chains. For example, arginine has a very polar ionized guanidino group that is generally solvent exposed. This group is linked to the C_α carbon, however, by a string of three hydrophobic methylene groups that are frequently buried. Furthermore, even large proteins have only a small proportion of their residues totally inaccessible to solvent.

Internal polar groups in proteins are mainly those that are present in the peptide backbone, which tend to be hydrogen bonded in secondary structure (14). Charged residues, even those paired in salt bridges with net neutrality, are comparatively rare in the interiors of proteins; about 96% of charged groups are located at the surface. Paradoxically, water molecules are often accommodated within the interiors of proteins, but these fill rare cavities and invariably participate in hydrogen bonding of polar groups. Such incorporated water molecules are often conserved through evolution, indicating that they are important parts of the protein structure.

3.3 Secondary structure

Taking all proteins of known structure together, 89% of residues are involved in secondary structure; about 31% of them occur in α-helices, 28% in β-sheets, and 30% in loops and turns. In general, α-helices and β-sheets occur in short segments, their lengths limited by the diameter of the domain, and it is often difficult to discern exactly where they start and stop. The average α-helix is 17 Å long and contains 11 residues, which corresponds to only three turns. The three residues at either end are each involved in only one hydrogen bond in an ideal α-helix. Individual β-strands are, on average, 20 Å long, which corresponds to 6.5 residues; they usually consist of from three to ten residues.

The idealized α-helix and β-sheet conformations described in Chapter 1 (*Figures 1.4* and *1.5*) are often distorted in proteins. For example, the terminal residues of α-helices often adopt a single turn of 3_{10} helix (Chapter 1, Section 5.2.2) (16). Most β-sheets in folded proteins have a right-handed twist, making the φ and ψ bond angles more positive than in the ideal case (17). Further distortions are common in β-sheets that consist of both parallel and anti-parallel strands, which have somewhat different structural requirements, and at the edges of β-sheets, where the presence of an extra residue sometimes disrupts the hydrogen bonding pattern to produce a β-bulge (*Figure 2.1b*).

3.4 Loops and turns

Most loops and turns occur at the surfaces of proteins and contain relatively polar residues (18–21). Loops containing fewer than about five amino acids can be classified according to their structure, but longer loops are more variable, and often flexible, without a well-defined structure. The most common type of loop links anti-parallel β-strands and is often referred to as a β-*turn* or, when the strands are adjacent in the β-sheet, as a β-*hairpin*. Most β-turns and β-hairpins are defined as involving the two residues not participating in the hydrogen bonding of the β-sheet plus the two flanking residues. Different types of β-turn are distinguished by the dihedral bond angles adopted by the central residues of the turn and the position of the hydrogen bond that usually occurs across the turn (*Figure 2.1*). For some of these turns only specific amino acids are allowed at particular positions, because of the dihedral bond angles that they can adopt. Glycine and proline residues are frequently found in turns because of their special conformational properties, and the sequence Gly-Pro is particularly common. Proline residues in turns are occasionally preceded by a *cis* peptide bond, which produces an abrupt reversal of the polypeptide direction.

3.5 Disulphide bonds

Disulphide bonds are covalent cross-links between two cysteine residues of the same or different peptide chains. They are most frequently found in proteins that are extracellular (secreted) and function under more oxidizing conditions where they are expected to provide the greatest stabilizing contribution. Disulphide bonds are formed during the process of folding (Chapter 3, Section 5.2) and require that the sulphur atoms of the participating cysteine residues be held in a closely-defined conformation with respect to each other (*Figure 2.2*) by the tertiary structure of the protein. For these reasons they are considered here as an integral part of the tertiary structure.

3.6 Domains

Large proteins are generally found to be made up of a number of globular units, called *domains*, linked together by short lengths of peptide chain. They can be independent units that are simply tethered by the linking polypeptide chain, or they can interact structurally to varying degrees. There is no clear cut distinction as to what constitutes a domain in a protein, and a variety of definitions are used. Each domain frequently includes only residues that are consecutive in the peptide chain, and such domains can often be separated by proteolytic cleavage of the linking peptide chain. Well-defined domains are often considered as *autonomous folding units*, because such individual domains usually fold independently of each other. However, other domains are not so structurally independent.

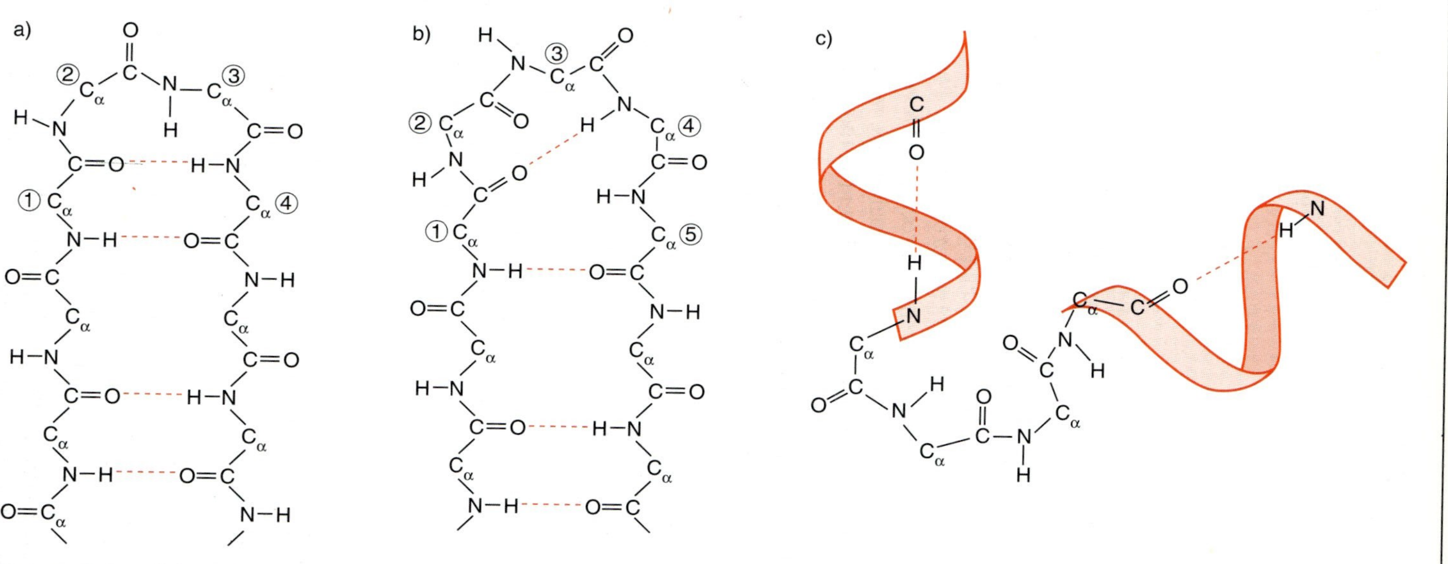

Figure 2.1. Schematic representations of some turn structures, in which the hydrogen bonds are shown as the orange dotted lines. (**a**) A β-turn in which residues one to four form the turn. (**b**) A β-turn followed by a β-bulge (see Section 3.3). Residues one to four form the turn and residue five is the extra 'bulge' residue. (**c**) A 2:4 turn linking two α-helices. This type of turn allows the helices to lie at angles of 30° to 60° with respect to each other and is similar to the type of turn found in the helix–turn–helix DNA binding motifs (see Chapter 4, Section 7.1). Many other types of turn structure have been classified. Adapted from Thornton, J.M., Sibanda, B.L., Edwards, M.S., and Barlow, D.J. (1988) *BioEssays*, **8**, 63.

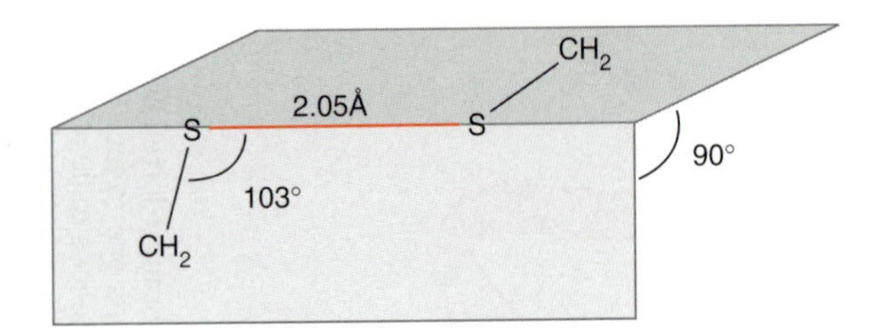

Figure 2.2. Optimal geometry of a disulphide bond. The two possible torsion angles of +90° (right-handed) and −90° (left-handed) are equally favoured. Other bond rotations are unfavourable to varying extents.

Proteins that have probably arisen fairly recently in evolutionary terms are frequently observed to be mosaics, apparently built up by linking together two or more domains that had been duplicated from other proteins (22). These proteins appear to have been generated genetically by a process of *domain shuffling*. Domains are often coded at the gene level by individual exons, which are those segments of DNA, separated by introns, that are present in the mature mRNA. At least some proteins have been generated genetically by shuffling of exons, but the generality of this phenomenon is still being debated (23).

3.7 Supersecondary structure and folding patterns

3.7.1 Chain topology

At the atomic level, protein structures appear extremely complex and unique. If, however, attention is focused on the pattern of secondary structure interactions within individual domains, it becomes apparent that many protein structures have topological similarities and can be broadly described using a finite number of folding patterns, which describe both the types of secondary structure present and their topological relationship. This is probably a consequence of all proteins containing primarily α-helices, β-strands, and turns and there being a limited number of ways in which these can be linked together. The following topological rules are obeyed in most cases (24,25).

1. Elements of secondary structure that are sequential in the primary structure are frequently adjacent in the tertiary structure and tend to pack in an anti-parallel manner (*Figure 2.3*).
2. Connections between secondary structure elements do not cross each other or make knots in the chain.
3. The connections between β-strands in most of the common β-X-β units, where the β-strands are parallel in the same β-sheet linked together by X, which can be an α-helix, a strand of a different sheet, or an extended piece of polypeptide chain, are right-handed (*Figure 2.3*).

Preferred packing patterns of secondary structure, such as the β-α-β unit, are known as *supersecondary structures*. They are frequently organized into common

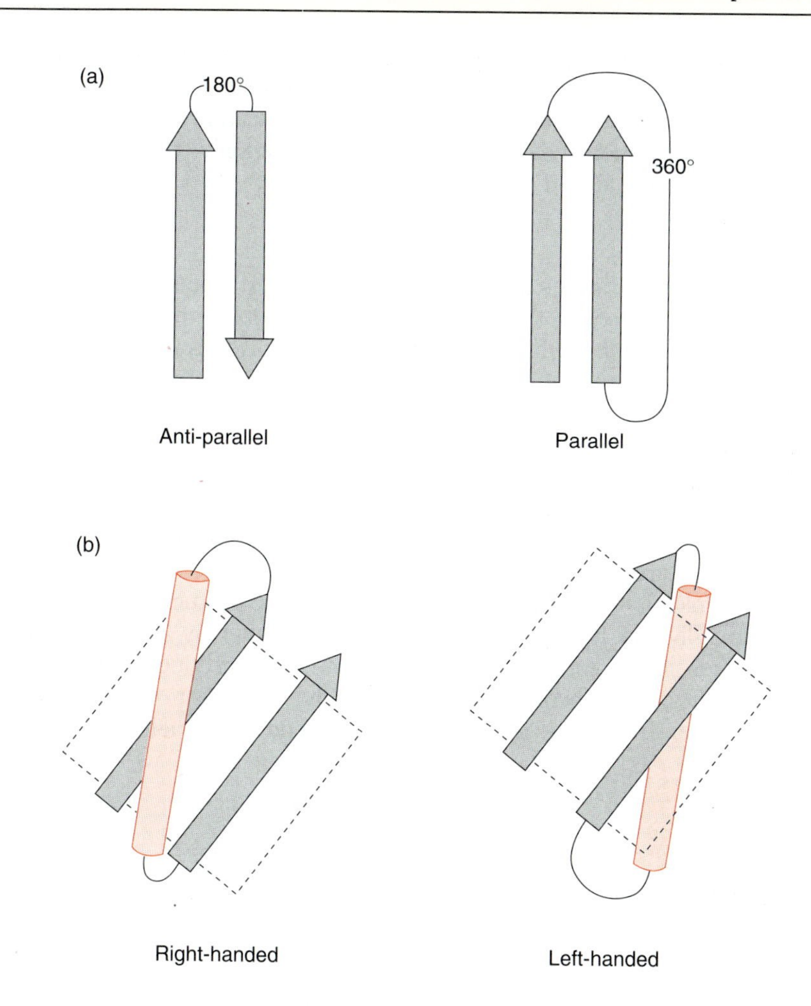

Figure 2.3. Preferences for certain topologies in protein structures. (**a**) Elements of secondary structure that are adjacent in the primary structure tend to pack together in an anti-parallel manner. This allows shorter loop lengths and reduces the extent to which the connecting loop must bend. (**b**) In β-X-β units the right-handed twist of the β-sheet (Section 3.3) favours right-handed connection of structural elements rather than the left-handed connection.

folding motifs or 'folds', such as the *Rossmann fold,* which is two linked β-α-β-α-β units (see Chapter 4, Section 6).

The classification of protein folds is somewhat arbitrary, and different overlapping classification schemes have been developed. Highly schematized representations of protein structure are often used to convey the relationship between secondary structure elements in protein folds. Protein structures are simplified and idealized by depicting only the path of the polypeptide chain as a ribbon, with α-helices as coils or cylinders and β-strands as arrows (26). While these simple diagrams convey many important features, and emphasize similarities between structures, they are idealized and usually omit the distortions present in the secondary structure of proteins. A complete classification system of protein folds cannot be examined in detail here, but some commonly encountered folding patterns (27) are considered in the following sections.

3.7.2 α-Proteins

The packing of two α-helices against each other is most frequently described in terms of *ridges into grooves*; the side chains of each helix form a series of ridges that are separated from each other by grooves. The ridges of one helix interdigitate with the grooves of the other when the helices interact closely. This interdigitation is most favourable when the axes of the helices are inclined at an angle of about +20° or −50° (defined in *Figure 2.4*) to each other. The coiled-coil (see Chapter 1, Section 6.3) is a similar form of this packing where the interaction between the helices is maintained over long distances by distortion and supercoiling of the individual helices. By comparison, the assemblies of helices in globular proteins are much shorter in length.

A common folding motif found in a number of diverse proteins is the *four-helix bundle*. The individual helices are inclined to one another by about 18° and are linked together in an anti-parallel arrangement by short loops. The helices diverge from the point at which they cross, which is the point of their maximum interaction. When this crossing point is near one end of the helices, an interior

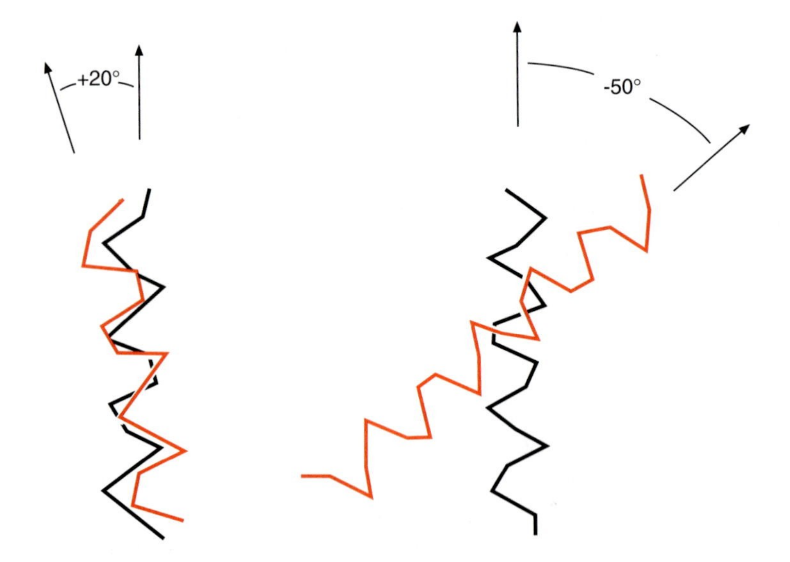

Figure 2.4. The preferred packing orientations of α-helices, with cross-over angles of +20° or −50°. These orientations maximize the interdigitation of ridges into grooves (Section 3.7.2). Figure kindly provided by C.Chothia.

cavity is formed at the other end; this has been adapted in proteins such as myohaemerythrin and cytochrome *c′* (*Figure 2.5*) to accommodate a prosthetic group, binuclear iron and haem, respectively. The lengths of the helices are from 15 to 24 residues, so they are somewhat longer than a typical α-helix. The anti-parallel arrangement of the helices may result from the short lengths of the linking loops and may be favoured by the relative orientation of the helix dipoles.

Assemblies of more than four α-helices are more complex structures (25), but the helices usually pack together to form compact aggregates around a central hydrophobic core.

3.7.3 β-Proteins

β-Strands form β-sheets by adopting a number of chain topologies (*Figure 2.6*) that follow the general criteria described in Section 3.7.1 (17). The adjacent strands are usually anti-parallel and linked by turns or loops of varying lengths. The β-hairpin and the β-meander are particularly common folding patterns, in which two or more anti-parallel β-strands that are sequential in the primary

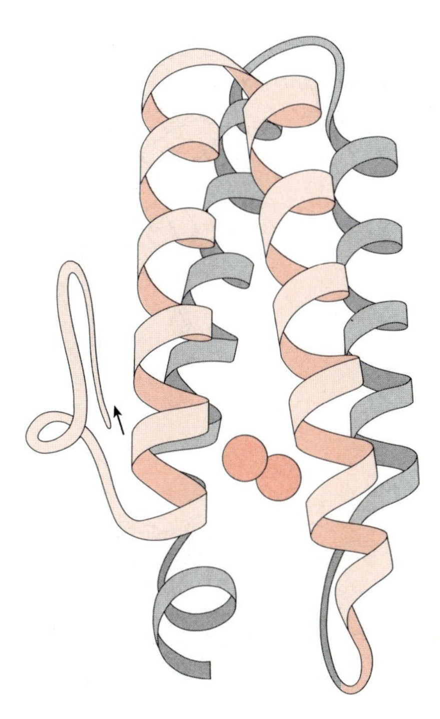

Myohaemerythrin

Figure 2.5. Schematic representation of the four-helix bundle protein myohaemerythrin. The central cavity contains binuclear iron. Reproduced with permission from Richardson,J.S. (1981) *Adv. Protein Chem.*, **34**, 167.

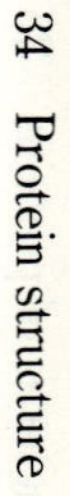

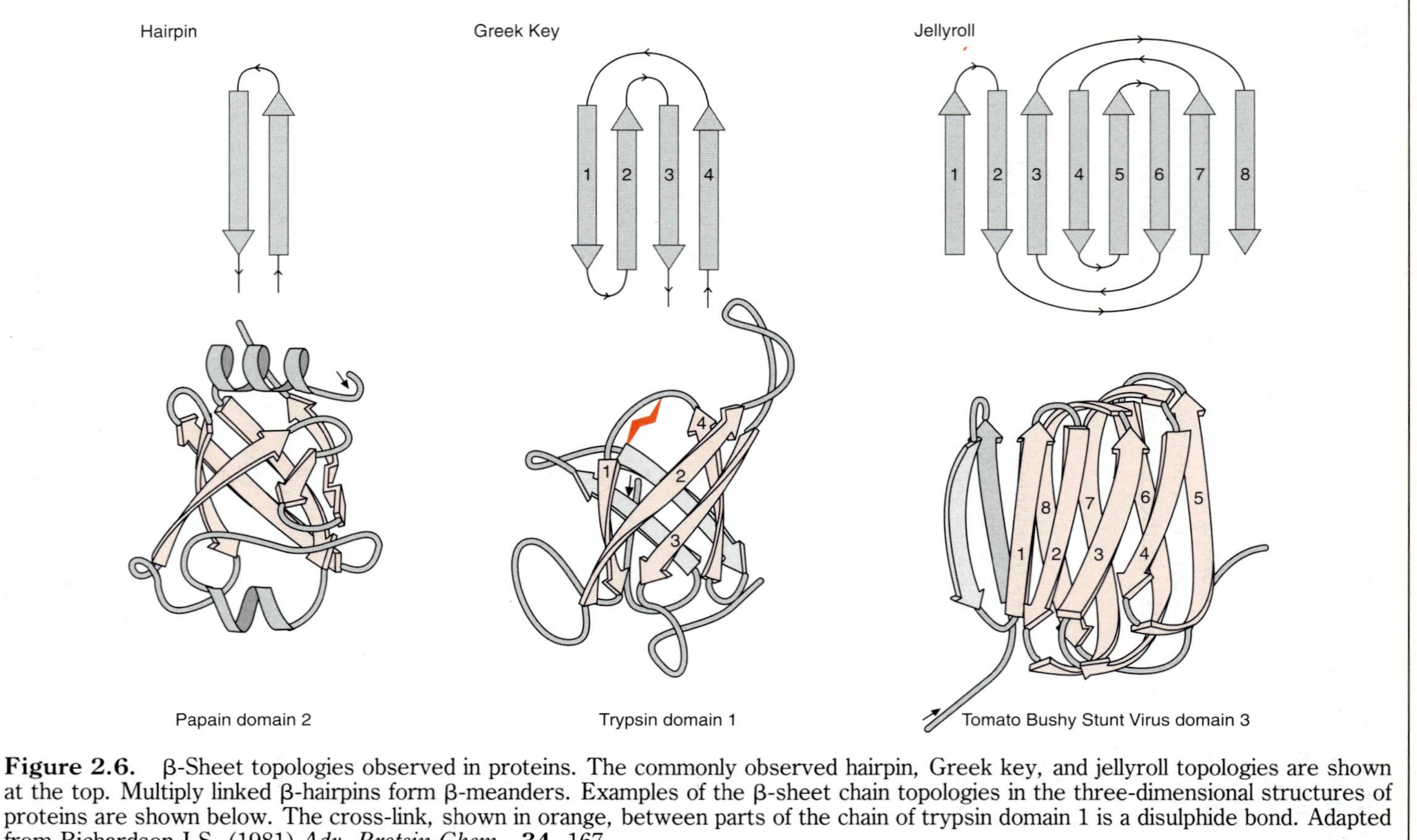

Figure 2.6. β-Sheet topologies observed in proteins. The commonly observed hairpin, Greek key, and jellyroll topologies are shown at the top. Multiply linked β-hairpins form β-meanders. Examples of the β-sheet chain topologies in the three-dimensional structures of proteins are shown below. The cross-link, shown in orange, between parts of the chain of trypsin domain 1 is a disulphide bond. Adapted from Richardson,J.S. (1981) *Adv. Protein Chem.*, **34**, 167.

structure are adjacent in the same β-sheet and linked by β-turns. The *Greek key* and *jellyroll* motifs differ in their connectivities between the strands, with longer loops connecting some of them.

β-Sheets often form layered or barrel structures (*Figure 2.6*). The strands of the sheets form the ribs of the barrel, and around the girth is the hydrogen bonding network. Non-polar side chains face inwards to form the hydrophobic core. In reality, the idealized barrel structure is often distorted and less evident than, for example, in $(\alpha/\beta)_8$ barrel structures described in Section 3.7.4. The β-barrel may not be completely closed, because of missing hydrogen bonds, or its cross section may be flattened to form an ellipse. In extreme cases, it is perhaps more appropriate to describe the structure as a β-sandwich, with a filling of non-polar side chains between two β-sheets.

3.7.4 α/β Proteins

In α/β proteins, α-helices and β-sheets alternate along the peptide chain. Small α/β proteins are usually simple two-layer structures with the helices packing against the β-sheet. For larger α/β proteins, there are two predominant folding patterns, exemplified by the structure of triose phosphate isomerase (TIM) and lactate dehydrogenase (LDH).

The $(\alpha/\beta)_8$ barrel structure of TIM is a common folding motif that has been found in at least 16 other proteins. Eight parallel strands of β-sheet coil around sequentially to form a central β-barrel. Concentric with the barrel and linking the β-strands are the parallel segments of α-helices (*Figure 2.7*).

In many proteins, including LDH, a central predominantly parallel twisted β-sheet is surrounded by an array of α-helices or loops. This $(\beta\text{-}\alpha\text{-}\beta\text{-}\alpha\text{-}\beta)_2$ unit, the Rossman fold, is particularly important in nucleotide binding proteins (Chapter 4, Section 6 and *Figure 4.2*).

3.8 Quaternary structure

Many proteins are made up of multiple polypeptide chains or subunits, which may be identical or different. The subunits are often essentially independently folded structures that interact because they have surfaces that are complementary in shape and physical interactions (28). The interactions between subunits involved in quaternary structure are not intrinsically different from those that occur within the subunits. Non-polar interactions tend to predominate at the centre of the interface, and hydrogen bonds and salt bridges at the periphery. In some dimers of proteins with β-sheets, hydrogen bonds between the β-sheets of the two subunits merge them into a single sheet. In other cases, independent domains can be considered subunits that are linked by a segment of polypeptide chain, so domain structure can be considered a form of quaternary structure.

The strength of the interaction between subunits can vary widely; some quaternary structures are extremely stable, so as to be difficult to dissociate, whereas others are readily reversible. In some cases the interactions are so weak that they are of dubious physiological significance. The tightness of binding

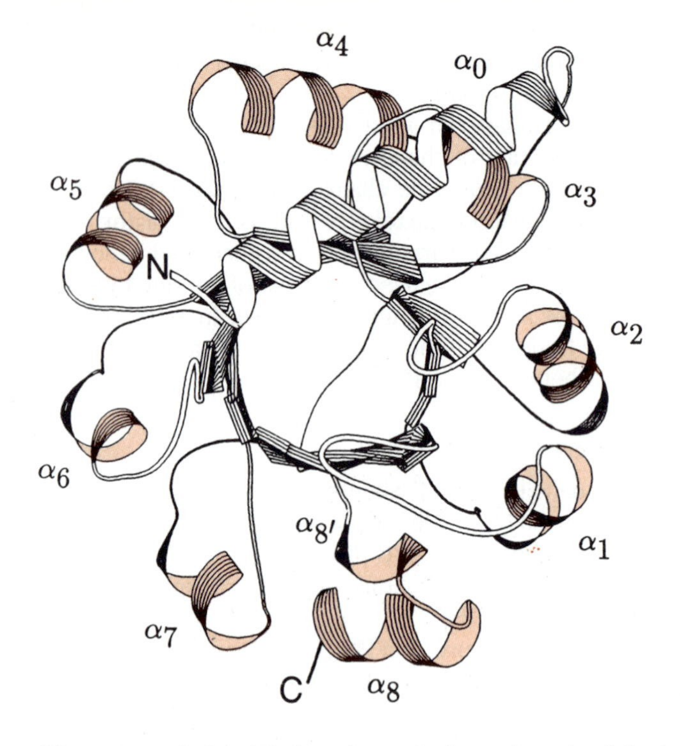

Figure 2.7. View of a typical $(\alpha/\beta)_8$ barrel protein down the axis of the barrel. The central β-strands all point out of the plane of the paper, whereas the α-helices project downwards. The particular structure shown is of indole glycerol phosphate synthase. It has an additional α-helix (α_0) at the N-terminus and a short helical segment ($\alpha_8{}'$) preceding helix 8. Reproduced with permission from Nermann, T. and Kirschner, K. (1990) *Protein Eng.*, **4**, 137.

is determined by the number and strengths of the contacts made between the subunits, which must be extensive to overcome the entropic advantage of the independent subunits and ensure stable association. Often there is a balance between the stability of the interaction and the need for some subunits to move with respect to each other as part of their function, for example in allosteric proteins (see Chapter 4, Section 8.1).

Depending upon the mode of association, there may be a fixed number of subunits, frequently two or four, or a variable number. Interactions between identical subunits are either *isologous* or *heterologous* (29). Isologous interactions involve the same surfaces on each monomer, which must be self-complementary, whereas heterologous interactions utilize different surfaces. Isologous interactions are limited in the first instance to producing dimers, but additional sets of isologous interactions can produce tetramers or higher-order structures. Isologous interactions produce structures with fixed numbers of subunits.

Heterologous interactions between monomers have no inherent limitation to the size of the aggregates they produce, and they frequently lead to indefinite polymerization. Indeed, this is the mechanism by which some elongated structural polymers, such as actin, are formed. With certain geometries of heterologous association, however, the outcome is a closed ring that contains a fixed number of subunits, depending upon the precise geometry; examples that have from three to 17 subunits are known. More complex interactions are apparent in closed spherical structures, such as those of icosahedral spherical viruses, which are built from 180 identical polypeptide chains. The identical subunits are only quasi-equivalent, because the formation of such a structure requires that they be placed in three different positions. The subunits use flexibility between domains to achieve the three types of structures required (30).

4. Flexibility in protein structures

The usual representations of protein structures derived from crystallographic data tend to give the impression that proteins have static structures. It must be kept in mind that there are varying degrees of flexibility within proteins, in that the various parts can adopt a range of closely-related conformations. This is often apparent crystallographically as a smearing of the electron density and is measured as the *B* value (31). In an ideal situation, where it is due solely to intramolecular flexibility, the *B* value would be proportional to the square of the vibration amplitude of the atoms in the molecule. In reality, however, it also reflects the static disorder of molecules in the crystal lattice.

At the atomic level, atoms in proteins are probably vibrating by fluctuations of bond lengths and angles at rates similar to those in any other molecule. Considering conformational changes due to bond rotations, the most mobile atoms are at the surface of the protein, where their flexibilities are like those of small molecules. Indeed, some atoms at the surface may be so mobile that their electron density is smeared out sufficiently that they are not visible in electron density maps. At other sites, however, the flexibility is curtailed to varying extents by the close packing of atoms; movement of one atom requires the concerted movement of other atoms.

Even within the interior of a protein, side chains of amino acids can be relatively mobile (32). NMR shows that the aromatic rings of many phenylalanine and tyrosine residues rotate 180° about the C_β–C_γ bond at least 10^4 times per second, because their δ and ε hydrogen atoms give averaged resonances. This occurs even when the side chains are totally buried in the interior. Such ring flipping of interior residues provides evidence for general flexibility in the protein structure, as other atoms in the structure must move to make it possible. Few phenylalanine and tyrosine residues in proteins flip less frequently, but tryptophan residues are much more fixed; the asymmetry of the larger indole side chain requires complete flips of 360°, not just half-flips of 180°.

Further evidence for the general flexibility of protein structure has come from the observation that the hydrogen atoms of all protein amide groups, even those that are buried, exchange with different isotopes (^{1}H, ^{2}H, and ^{3}H) in the solvent, albeit at low rates (33). The molecular mechanism of hydrogen exchange is not known in any case. One model is that local unfolding or 'breathing' transiently exposes the buried residues to the bulk solvent. In an alternative model, occasional 'pores' open up in the protein, through which water molecules diffuse into the interior to exchange with the amide. Neither model is totally consistent with all the experimental observations, so several different mechanisms probably operate.

5. Evolutionarily related proteins

New proteins have usually evolved by duplication of an existing gene and mutational divergence of the two new genes and their protein products. In this way, the original function is retained by one gene while the other provides a folded structure that can be altered in function by mutation. Proteins with similar primary sequences always adopt similar three-dimensional structures; the three-dimensional structure of a protein seems to have been more conserved during evolution than has the primary structure (34,35). Many proteins differ dramatically in their amino acid sequences, yet adopt the same overall folded conformation. For example, the members of the globin family have only two residues in common, yet all adopt very similar folded structures. Even proteins not detectably homologous are occasionally found to adopt similar conformations. These similar structures illustrate that there is considerable degeneracy in the way the amino acid sequence information specifies the structure of a protein.

The most frequent kinds of amino acid replacements during protein evolution are those that conserve the type of side chain, for example arginine for lysine, aspartic acid for glutamic acid, phenylalanine for tyrosine, etc. Functionally important residues, such as the haem-binding histidine residue of the globins and the catalytic residues of enzymes, are the most highly conserved. In general, the most frequent sites of mutations are non-functional residues at the surfaces of proteins, especially those in loops; insertions and deletions occur there most frequently and often lead to local alterations in conformation. Disulphide bonds have been inserted or deleted during evolution, and usually both cysteine residues are replaced or inserted. Mutations in the hydrophobic cores of proteins occur comparatively rarely, presumably because they tend to disrupt critical packing interactions that are required for conformational stability (Chapter 3, Section 4.3). Individual mutations that have occurred during evolution rarely alter the protein structure dramatically, but their cumulative effect can be substantial, changing the lengths of secondary structure elements and their positions relative to each other, or even deleting them entirely. The vast majority of evolutionary change at the molecular level seems to be functionally neutral; natural selection has been primarily negative in selecting against mutations that produce functional change. Only in a few instances is there evidence for positive selection for functional change (36).

When no sequence homology is apparent between proteins of similar structure, and a phylogenetic tree linking them cannot be traced, there are two possibilities. Either the proteins have diverged from an ancient common ancestor to such an extent that their common origin is no longer apparent in their primary structures, or their similar folded structures arose independently through *convergent evolution* (37). Convergence is plausible, but difficult to prove, and there are no established examples. Convergent evolution has occurred to develop the trypsin and subtilisin families of proteases, which utilize similar catalytic mechanisms, but have very different three-dimensional structures. The TIM $(\alpha/\beta)_8$ barrel has been proposed as a common protein fold that has arisen by convergent evolution, since many of the 17 known examples of this fold have no sequence homology and only slight functional similarities (38). The $(\alpha/\beta)_8$ barrel also represents one of the simplest structures an α/β protein can adopt that fulfils the general topology rules described in Section 3.7.1; consequently, this structure is a plausible candidate for having arisen independently in different proteins (38).

6. Predicting protein structure

The three-dimensional structure of a protein is determined by its amino acid sequence, so it should be possible to predict the folded conformation directly from knowledge of this sequence. The *ab initio* calculation of protein structure from the amino acid sequence is not feasible at present. There are, however, a number of procedures available for predicting secondary structure, which rely on the slight statistical preferences that individual amino acids show for adopting or terminating α-helical, β-sheet, or turn conformations (39,40). The most likely secondary structure is evaluated from the balance of these preferences over a number of residues adjacent in the sequence. Refinements of this basic method take into account the observation that particular residues are frequently found at certain positions in secondary structure (Chapter 1, Section 5.2).

The procedures for predicting helix, sheet, reverse turn, or irregular structures have an accuracy of only about 60% (41). The primary source of error is probably that local sequences are not the sole determinant of secondary structure (42). At least some of the secondary structure preferences exhibited by particular residues may not be due to their inherent conformational properties but to their overall role in the tertiary structure of proteins; for example, β-sheets tend to be buried in protein structures, so amino acids with non-polar side chains may predominate in β-strands simply because of their tendency to be buried, not their intrinsic preference to form particular secondary structures.

Tertiary structures can be predicted from the amino acid sequence at the present time only if the structure of a homologous protein is known. In this case, the amino acid differences in the new sequence can be built into the known structure to generate a model for the new protein; its plausibility is inversely proportional to the number of amino acid changes, especially insertions and deletions. A more sensitive approach to searching for sequences that adopt the same conformation is to define the critical residues required for that particular structure, which are often known as a *tertiary template* (43). This is defined in the most general terms, such as whether polar/non-polar, charged/uncharged, small/large residues are required at particular positions in the structure. The unknown sequence is then compared to the general template to predict whether the protein will adopt that folded structure. Another method considers the likelihood that each amino acid residue of a primary structure would occur in the corresponding position in a three-dimensional structure (44).

7. Protein design

The converse problem to predicting protein structure is to design proteins *de novo*. Factors known to stabilize secondary structure elements and favour packing interactions between them have been incorporated into a number of protein designs, notably to make four-helix bundle proteins (45), some simple β-sheet structures, and an $(\alpha/\beta)_8$ barrel (46). Simple functions such as metal binding have also been successfully incorporated (47). The design of the four-helix bundle illustrates a number of principles already discussed (*Figure 2.8*).

Design of more complex structures awaits a more complete understanding of how non-covalent forces in proteins specify and stabilize the overall three-dimensional structure.

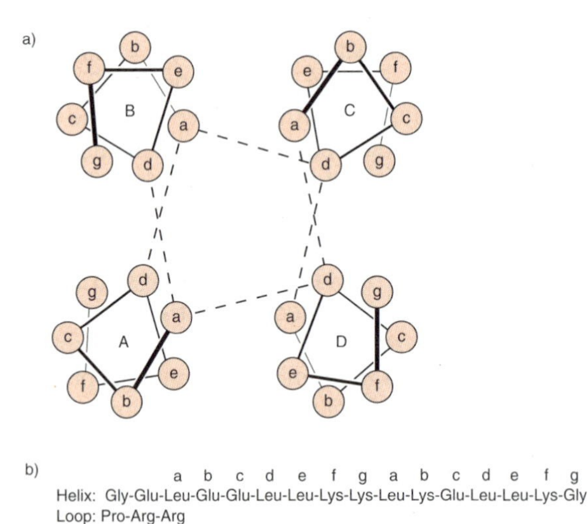

Figure 2.8. (**a**) Idealized view of a four-helix bundle protein in which residues a and d in each helix interact by non-polar interactions. Reproduced with permission from Cohen, C. and Parry, D.A.D. (1990) *Proteins: Struct. Funct. Genet*, **7**, 1. (**b**) Sequence of a designed four-helix bundle protein. Acidic and basic residues at the N and C termini, respectively, should stabilize the helix macro dipole. The distribution of the residues was chosen to favour formation of an amphipathic helix, with leucine residues occurring on one face and the glutamic acid and lysine residues on the other. Ion pairs were designed to form between the charged residues along one face, whilst the non-polar surfaces of the helices pack together (see Chapter 1, Section 5.2.1).

8. Further reading

Blundell,T.L. and Johnson,L.N. (1976) *Protein Crystallography.* Academic Press, New York.

Brandén,C. and Tooze,J. (1991) *Introduction to Protein Structure.* Garland Publishing, New York.

Chothia,C. (1984) Principles that determine the structure of proteins. *Annu. Rev. Biochem.*, **53**, 537.

Fletterick,R.J., Schroer,T., and Matela,R.J. (1985) *Molecular Structure: Macromolecules in Three Dimensions.* Blackwell Scientific, Oxford.

Lesk,A. (1991) *Protein Architecture: a Practical Approach.* Oxford University Press, Oxford.

McCammon,A. and Harvey,S.C. (1987) *Dynamics of Proteins and Nucleic Acids.* Cambridge University Press, New York.

McPherson,A. (1982) *Preparation and Analysis of Protein Crystals.* John Wiley, New York.

Richardson,J.S. (1981) The anatomy and taxonomy of protein structure. *Adv. Protein Chem.*, **34**, 167.

Schulz,G.E. and Schirmer,R.H. (1979) *Principles of Protein Structure.* Springer, New York.

Wüthrich,K. (1986) *NMR of Proteins and Nucleic Acids.* John Wiley, New York.

9. References

1. Johnson,W.C, Jr. (1990) *Proteins: Struct. Funct. Genet.*, **7**, 205.
2. Manning,M.C. and Woody,R.W. (1989) *Biochemistry*, **28**, 8609.
3. Williams,R.W. (1986) *Methods Enzymol.*, **130**, 311.
4. Weber,P.C. (1991) *Adv. Protein Chem.*, **41**, 1.
5. Hajdu,J. and Johnson,L.N. (1990) *Biochemistry*, **29**, 1669.
6. Hendrickson,W.A., Horton,J.R., and LeMaster,D.M. (1990) *EMBO J.*, **9**, 1665.
7. Lawrence,M.C. (1991) *Q. Rev. Biophys*, **24**, 399.
8. Wlodawer,A. (1982) *Prog. Biophys. Mol. Biol.*, **40**, 115.
9. Kossiakoff,A.A. (1985) *Annu. Rev. Biochem*, **54**, 1195.
10. Wüthrich,K. (1989) *Acc. Chem. Res.*, **22**, 36.
11. Clore,G.M. and Gronenborn,A.M. (1991) *Annu. Rev. Biophys. Biophys. Chem.*, **20**, 29.
12. Braun,W. (1987) *Q. Rev. Biophys.*, **19**, 115.
13. Herzberg,O. and Moult,J. (1991) *Proteins: Struct. Funct. Genet.*, **11**, 223.
14. Baker,E.N. and Hubbard,R.E. (1984) *Prog. Biophys. Mol. Biol.*, **44**, 97.
15. Richards,F.M. (1977) *Annu. Rev. Biophys. Bioeng.*, **6**, 151.
16. Barlow,D.J. and Thornton,J.M. (1988) *J. Mol. Biol.*, **201**, 601.
17. Salemme,F.R. (1983) *Prog. Biophys. Mol. Biol.*, **42**, 95.
18. Venkatachalam,C.M. (1968) *Biopolymers*, **6**, 1425.
19. Milner-White,E.J. and Poet,R. (1987) *Trends Biochem. Sci.*, **12**, 189.
20. Rose,G.D., Gierasch,L.M., and Smith,J.A. (1985) *Adv. Protein Chem.*, **37**, 1.
21. Sibanda,B.L., Blundell,T.L., and Thornton,J.M. (1989) *J. Mol. Biol.*, **206**, 759.
22. Doolittle,R.F. (1985) *Trends Biochem. Sci.*, **10**, 233.
23. Blake,C.C.F. (1985) *Int. Rev. Cytol.*, **93**, 149.
24. Finkelstein,A.V. and Ptitsyn,O.B. (1987) *Prog. Biophys. Mol. Biol.*, **50**, 171.
25. Chothia,C. and Finkelstein,A.V. (1990) *Annu. Rev. Biochem.*, **59**, 1007.
26. Richardson,J.S. (1985) *Methods Enzymol.*, **115**, 349.
27. Levitt,M. and Chothia,C. (1976) *Nature*, **261**, 552.

28. Miller,S. (1989) *Protein Eng.*, **3**, 77.
29. Monod,J., Wyman,J., and Changeux,J.-P. (1965) *J. Mol. Biol.*, **12**, 88.
30. Liljas,L. (1986) *Prog. Biophys. Mol. Biol.*, **48**, 1.
31. Ringe,D. and Petsko,G.A. (1985) *Prog. Biophys. Mol. Biol.*, **45**, 197.
32. Wagner,G. (1983) *Q. Rev. Biophys.*, **16**, 1.
33. Englander,S.W. and Kallenbach,N.R. (1984) *Q. Rev. Biophys.*, **16**, 521.
34. Bajaj,M. and Blundell,T. (1984) *Annu. Rev. Biophys. Bioeng.*, **13**, 453.
35. Chothia,C. and Lesk,A.M. (1986) *EMBO J.*, **5**, 823.
36. Creighton,T.E. and Darby,N.J. (1989) *Trends Biochem. Sci.*, **14**, 319.
37. Ptitsyn,O.B. and Finkelstein,A.V. (1980) *Q. Rev. Biophys.*, **13**, 339.
38. Farber,G.K. and Petsko,G.A. (1990) *Trends Biochem. Sci.*, **15**, 228.
39. Levitt,M. (1978) *Biochemistry*, **17**, 4277.
40. Chou,P.Y. and Fasman,G.D. (1978) *Annu. Rev. Biochem.*, **47**, 251.
41. Schultz,G.E. (1988) *Annu. Rev. Biophys. Biophys. Chem.*, **17**, 1.
42. Kabsch,W. and Sander,C. (1984) *Proc. Natl. Acad. Sci. USA*, **81**, 1075.
43. Taylor,W.R. (1989) *Prog. Biophys. Mol. Biol.*, **54**, 159.
44. Bowie,J.U., Lüthy,R., and Eisenberg,D. (1991) *Science*, **253**, 164.
45. Regan,L. and DeGrado,W.F. (1988) *Science*, **241**, 976.
46. Goraj,K., Renard,A., and Martial,J.A. (1990) *Protein Eng.*, **3**, 259.
47. Regan,L. and Clarke,N.D. (1990) *Biochemistry*, **29**, 10878.

3

Proteins in solution and in membranes

1. Introduction

Proteins have very varied physical and functional properties that are primarily dependent upon their surfaces. In biological systems, proteins exist mainly in either aqueous or membrane environments. Aqueous solubility is favoured by the presence of polar groups on the protein surface, whereas the surfaces of membrane proteins are much less polar.

The surface of a protein is, of course, determined by its three-dimensional folded conformation. The folded conformation is usually labile, but can often be regained after unfolding; this makes it possible to study the physical basis of the stability of the folded state and the folding process.

2. Proteins in aqueous solution

2.1 Aqueous solubility

In general the more polar its surface, the more soluble a protein is likely to be. Protein solubility may be increased by polar modifications such as glycosylation and phosphorylation but such predictions cannot be made with certainty, however, and even proteins with polar surfaces will be insoluble if they bind tightly to other macromolecules and consequently aggregate. This occurs with a number of structural proteins that have polar surfaces but which interact with each other more avidly than with water. Examples of these proteins are the structural proteins of viruses (1).

Soluble globular proteins are surrounded by a layer of water molecules that have somewhat different properties from normal water, but which are generally in dynamic equilibrium with the bulk water (2,3). There is usually 0.3 g of tightly bound water per gram of protein, or nearly two molecules per amino acid residue. Some relatively fixed water molecules are observed in protein crystal structures, which are hydrogen bonded to two or more charged or polar groups on the surface. Other less tightly fixed water molecules participate in single hydrogen bonds to the protein and in hydrogen bonded networks of multiple water molecules around the protein. The ordered networks of water molecules that might be expected around solvent-exposed non-polar atoms (Chapter 1, Section 2.4) are generally not observed crystallographically, probably because they are not fixed relative to the non-polar surface (4).

Proteins tend to be least soluble at pH values near their isoelectric point, where they have no net charge. Low concentrations of any salt usually increase the solubility of a protein by decreasing its electrostatic free energy. However, protein solubility decreases at high salt concentrations to varying extents, depending upon the salt. This effect results from a number of factors, including the screening out of electrostatic repulsions that cause the solute molecules to repel each other. High concentrations of salts also increase the surface tension of water to varying extents, making it energetically less favourable to create cavities in the solvent in which the soluble protein molecules can be accommodated. This is the basis of the well-known *Hofmeister series* of salts. However, this effect can be overcome if the salt binds to the protein surface. The surfaces of folded and unfolded proteins differ, so salts that interact with them in different ways will also affect the stability of the folded conformation. In general, the effects of co-solvents on protein aqueous solubility and stability are now understood in terms of their effects on the properties of water and their interactions with the surfaces of folded and unfolded proteins (5).

As might be expected, unfolded or partially folded proteins are generally less soluble than the fully folded protein, because of the greater exposure of non-polar groups. This can hinder the refolding of proteins *in vitro* and necessitates the occurrence of special mechanisms *in vivo* that keep unfolded and partially folded proteins soluble (Section 5.3).

2.2 Ionization

The ionization tendencies of polar groups in proteins are frequently perturbed from their intrinsic pKs, because of their proximity to and interaction with other atoms in the folded structure (6). For example, the pK values of the histidine side chains in myoglobin vary from 5.5 to 8.1. Such perturbations often result from the proximity of other charged groups, participation in a hydrogen bond or salt bridge, or from being placed in a hydrophobic environment. The magnitudes of these effects are complex, as no simple equivalent to the dielectric constant exists for the structurally heterogeneous protein interior; quantitative analysis of electrostatic effects in proteins is only now becoming possible (7,8).

In a number of instances, perturbation of the ionization of side chains is of functional importance, especially in ligand binding and catalytic activity. For example, the active site of lysozyme places a glutamic acid residue in a non-polar environment, elevating its pK so that it remains protonated at the enzyme's optimum pH to serve as a proton donor in the hydrolysis of oligosaccharides. In thiol proteases, the pK of the active site cysteine is reduced by interaction with adjacent charged residues so that it ionizes more readily and is more reactive towards the substrate. Extreme perturbation of pK values occurs only when important for function, because most such perturbations by the folded conformation decrease its stability.

Many non-ionized histidine and tyrosine residues are buried in protein interiors and can ionize only if the protein unfolds. This is a major reason why so many proteins unfold at acid or alkaline pH.

3. Membrane proteins

Membrane proteins are of considerable biological importance but, because they have proved difficult to crystallize, progress has been slow in determining their detailed structures. The few detailed structures known show them to be similar in many respects to soluble proteins. The main difference is that their surfaces are covered with predominantly non-polar groups that interact favourably with the non-polar environment of the lipid membrane (9). Indeed, integral membrane proteins can often be identified from their primary structures by the occurrence of unbroken stretches of non-polar amino acid residues (10).

The membrane is a dynamic environment and proteins in it can move laterally in the lipid bilayer, although because of their generally large size and high density in the membrane (which may be up to 25% protein by volume) their mobility is limited. Proteins rotate about axes perpendicular to the membrane but do not flip about axes parallel to it, and so retain a constant vertical orientation.

Integral membrane proteins are embedded within the lipid bilayer, although portions of the protein may lie outside it (11). The exposed portions of the protein can frequently be identified by their accessibility to proteases and to chemical or antibody probes. Proteins may also be associated with membranes solely by their interaction with integral membrane proteins or by use of special non-polar membrane anchor segments; these proteins are otherwise comparable to soluble globular proteins.

X-ray structures are now available for two very similar bacterial photosynthesis reaction centres (12,13) and for the bacterial membrane channel protein, porin (14). The interiors of the photosynthetic reaction centre proteins are very similar to the interiors of soluble proteins. Overall, these membrane proteins are distinguished from soluble proteins only by their non-polar surfaces that interact with the hydrophobic portions of the membrane bilayer (9). This contrasts with earlier ideas that membrane proteins would simply be 'inside-out' soluble proteins, with polar groups on the inside, but further structures are needed to test the generality of this observation.

Polar residues are often accommodated within the membrane-buried portions of proteins, but the mechanism by which this occurs is unclear. These internal polar residues may be functionally important, for example, in ion channels.

Integral membrane proteins most frequently have a high α-helical content. In the photosynthetic reaction centre proteins, there are 11 tightly packed α-helices of 19–30 residues in length that span the membrane (*Figure 3.1*). Solvent-exposed loops and α-helices connect the transmembrane helical

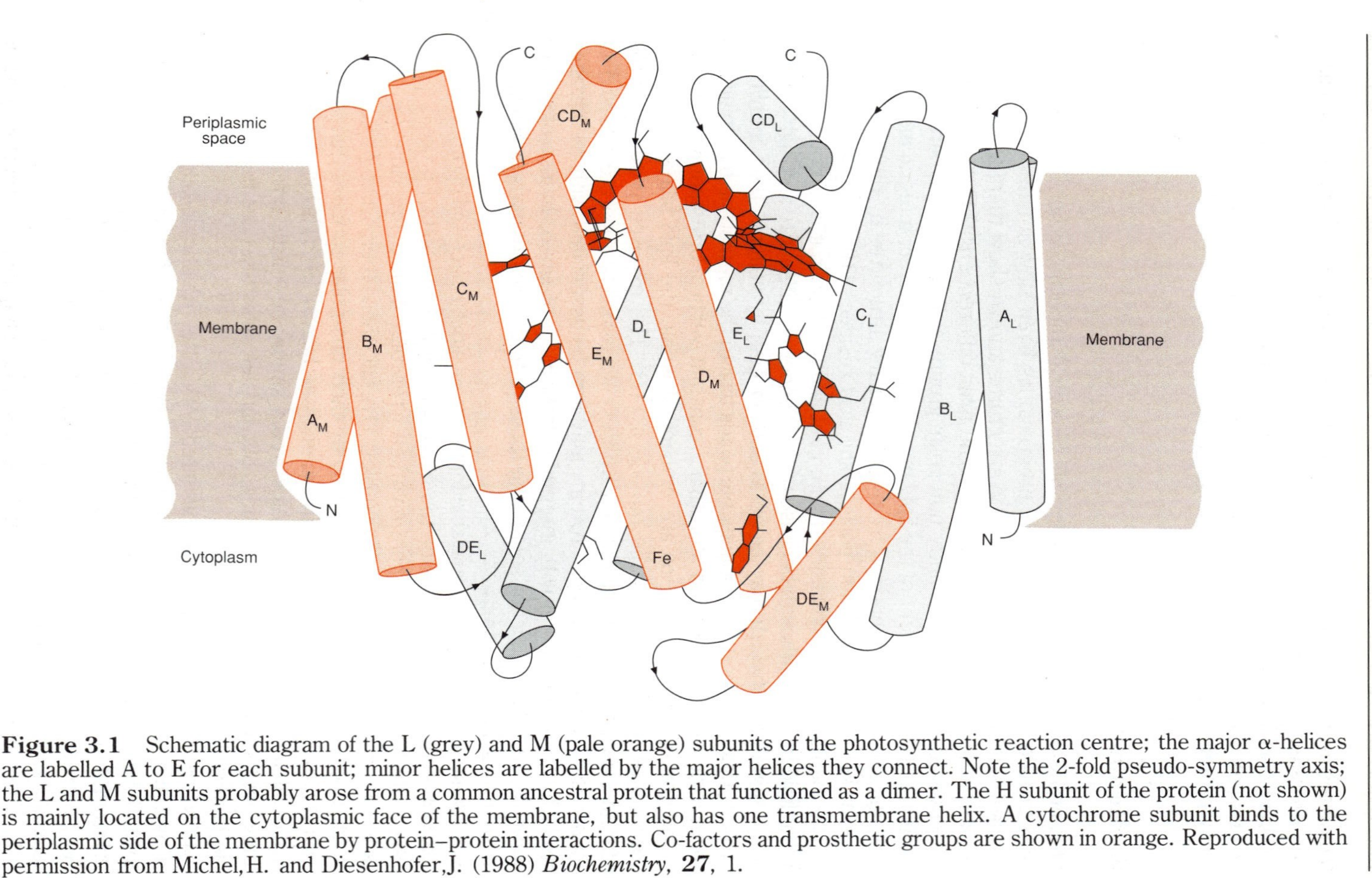

Figure 3.1 Schematic diagram of the L (grey) and M (pale orange) subunits of the photosynthetic reaction centre; the major α-helices are labelled A to E for each subunit; minor helices are labelled by the major helices they connect. Note the 2-fold pseudo-symmetry axis; the L and M subunits probably arose from a common ancestral protein that functioned as a dimer. The H subunit of the protein (not shown) is mainly located on the cytoplasmic face of the membrane, but also has one transmembrane helix. A cytochrome subunit binds to the periplasmic side of the membrane by protein–protein interactions. Co-factors and prosthetic groups are shown in orange. Reproduced with permission from Michel, H. and Diesenhofer, J. (1988) *Biochemistry*, **27**, 1.

segments at the membrane faces. The loops have positively charged side chains concentrated on the cytoplasmic side of the membrane and negatively charged side chains on the other side. The cytoplasmic side of the membrane carries a net negative charge as a result of ion pumping, which suggests that electrostatic effects may play a role in stabilizing the protein structure and correctly orienting it in the membrane (11,15).

α-Helices appear to be a general feature of many integral membrane proteins, but there is a notable exception, the protein porin (14). This is a 16-strand β-barrel, the inside of which forms a polar channel through the outer membrane of bacteria; the side chains on the outer surface of the barrel are non-polar, whereas the residues lining the channel have polar side chains (*Figure 3.2*). In the membrane, porin exists as a trimer. The interfaces between the subunits are predominantly polar, and they interact through salt bridges and hydrogen bonds.

4. Stability of the folded state

4.1 Unfolding

The structure of a protein can be disrupted by relatively minor alterations in pH or temperature, the addition of denaturing agents (e.g. urea) and, in specific cases, by reduction of disulphide bonds or removal of an essential prosthetic group or co-factor. These varying methods of denaturing proteins reflect the different types of interactions that stabilize the folded state. For example, denaturants such as urea and guanidinium salts interact favourably with both polar and non-polar surfaces (16), like those found in protein interiors (Chapter 2, Section 3.2); consequently the preferential interactions between groups that stabilize the folded protein are not so favourable. In contrast, pH effects on stability are primarily a result of the ionization of buried residues (see Section 2.2), plus some effects on a few salt bridges.

Loss of the folded structure of proteins can be readily followed by observing changes in absorption spectra, CD, or fluorescence spectra, or in the dimensions of the protein, which generally increase upon denaturation (17). Unfolding transitions monitored by any of these techniques for a single-domain protein generally approximate to a two-state situation in which only the fully folded and fully unfolded states are populated (*Figure 3.3a*). Larger proteins often give more complex transitions when their domains unfold individually. The domains may unfold independently of each other in two-state transitions, or there may be varying degrees of co-operativity between them.

The two-state nature of the unfolding transition of a protein domain results from the co-operative nature of the interactions that stabilize protein structure (Section 4.3). The stabilities of the interactions are so dependent upon each other that disruption of a very limited number of interactions tends to disrupt all of them.

The two-state nature of the unfolding transition simplifies the thermodynamic

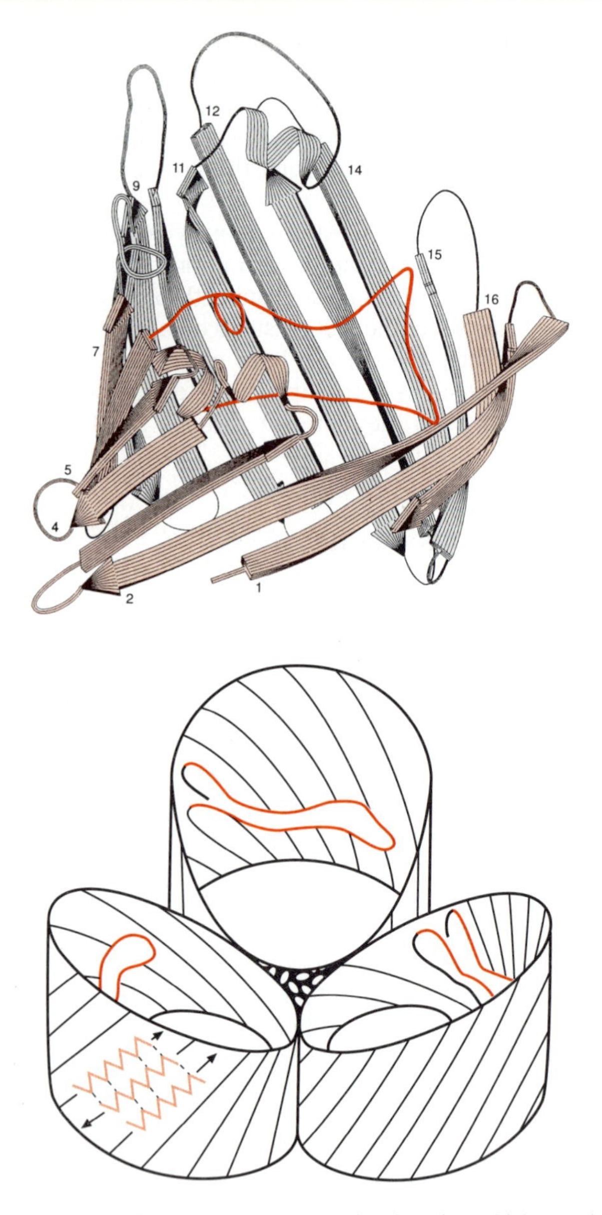

Figure 3.2. (**a**) Schematic view of one subunit of porin, which consists of 16 antiparallel β-strands, each 6–17 residues long, linked together by short lengths of α-helix or by loops to form a barrel. One loop of 44 residues (shown in orange) linking an α-helix after strand 5 to strand 6 protrudes into the barrel and restricts the channel. The bottom rim of the barrel faces the periplasm and is relatively flat, while the top rim with its longer and more irregular connections is slanted and also less regular. Reproduced, with permission from ref. 14. (**b**) Schematic representation of a porin trimer, with the channel-restricting loop shown in orange. Reproduced with permission from Schirmer, T. and Rosenbusch, J. P. (1991) *Curr. Opinion Struct. Biol.*, **1**, 539.

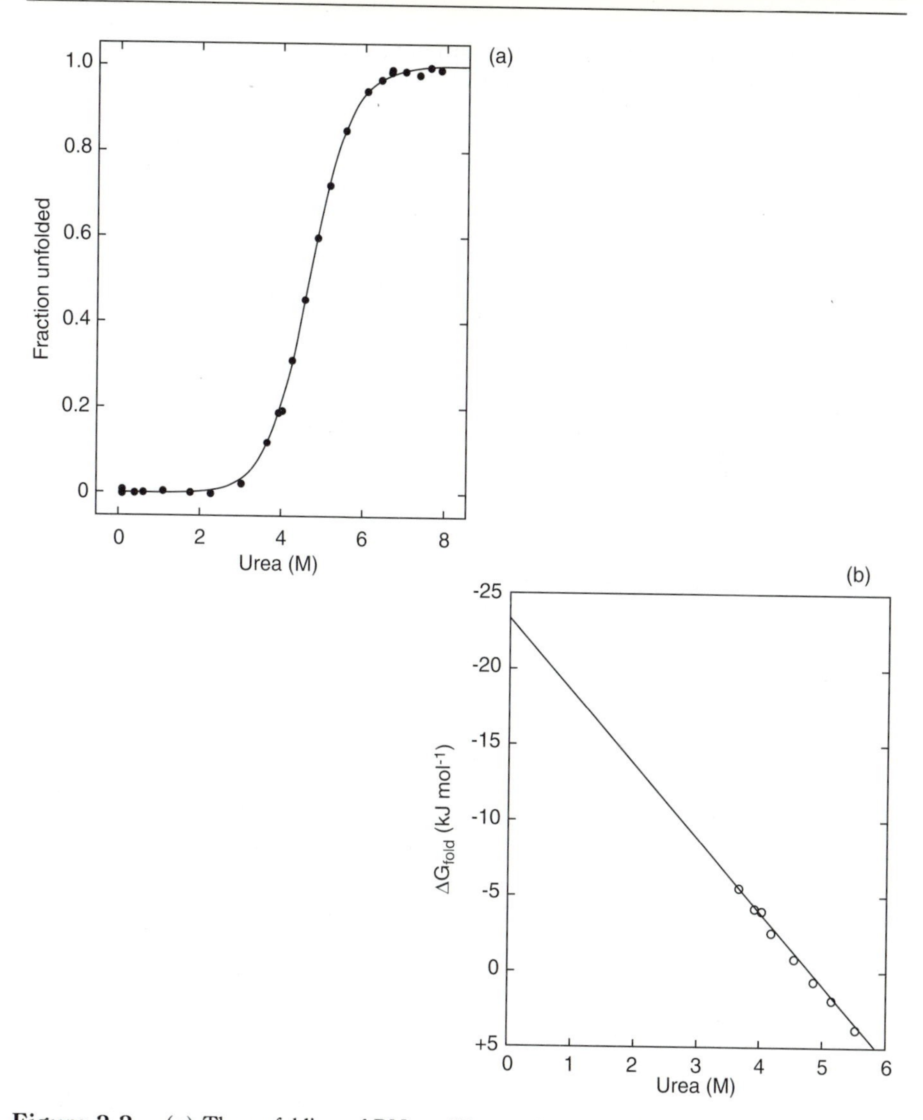

Figure 3.3. (**a**) The unfolding of RNase T1 as a function of urea concentration. The fraction of unfolded molecules was measured spectroscopically. Below 3 M urea, the folded (N) conformation predominates. A reversible, two-state ($N \leftrightarrow U$) unfolding transition occurs between 3 and 6 M urea, where the data give the relative stabilities of the U and N states as a function of urea concentration; such data can be extrapolated to higher and lower urea concentrations. The unfolded (U) state predominates above 6 M urea. Reproduced with permission from Pace, C.N., Shirley, B.A., and Thomson, J.A. (1989) In Creighton, T.E. (ed.), *Protein Structure: a Practical Approach*. IRL Press, Oxford, p.311. (**b**) The data from (a) are used to determine the value of $K_{\mathrm{eq,fold}}$ (equation 3.1) and hence ΔG_{fold} (equation 3.2). The plot of ΔG_{fold} as a function of urea concentration gives the value of $\Delta G_{\mathrm{fold,physiol}}$ as the y-axis intercept and the value of m as the gradient of the fitted line. The values determined here for RNase T1 are $\Delta G_{\mathrm{fold,physiol}} = -23.9$ kJ mol^{-1} and $m = +5.1$ kJ mol^{-1} M^{-1}.

analysis of unfolding. Any structural parameter can be used to determine the proportion of molecules in the folded (N) and unfolded (U) states at equilibrium throughout the transition region, which gives the equilibrium constant for folding, $K_{\mathrm{eq,fold}}$:

$$K_{\mathrm{eq,fold}} = [N]/[U] \tag{3.1}$$

The difference in free energy between the folded and the unfolded states, ΔG_{fold}, can be derived from $K_{\mathrm{eq,fold}}$ by

$$\Delta G_{\mathrm{fold}} = -\mathrm{R}T \log_e K_{\mathrm{eq,fold}} \tag{3.2}$$

The value of ΔG_{fold} is generally found to vary linearly as a function of the denaturant concentration with a proportionality factor (m) across the transition region. Consequently, extrapolation back to zero denaturant concentration or to physiological conditions gives an estimate of the value of ΔG_{fold} under those conditions, $\Delta G_{\mathrm{fold,physiol}}$ (*Figure 3.3b*; 18).

$$\Delta G_{\mathrm{fold}} = \Delta G_{\mathrm{fold,physiol}} + m[\mathrm{denaturant}] \tag{3.3}$$

Values of ΔG_{fold} measured in this and other ways vary from about -20 to -60 kJ mol^{-1}, indicating that folded proteins are only marginally stable. This corresponds to a $K_{\mathrm{eq,fold}}$ value of 10^4–10^7, so a small proportion of the molecules will be transiently unfolded even under optimal conditions.

Unfolding generally occurs without covalent alteration of the protein, unless disulphide bonds are reduced, and it is often reversible. The unfolded protein approximates to a random coil polypeptide chain (Chapter 1, Section 5.1.2) under strongly unfolding conditions, such as in 6 M guanidinium chloride. The diversity of chemical groups in a protein, however, makes it unlikely that the unfolded state can be a true random coil, and there are a number of indications that there may be varying degrees of preferential interactions between its various groups, depending upon the conditions. Any structure present in unfolded proteins is local, however, and the global co-operative interactions characteristic of the native state are absent.

The most common exception to two-state unfolding transitions is the occurrence of a stable, partially folded state, known as the *molten globule* (19). For reasons that are not clear, it is only stable with certain proteins under particular conditions. It is almost as compact as the fully folded protein and has a similar secondary structure content, but little or no tertiary structure. Its physical nature is not yet clear, but two extreme alternative models are that it is either an expanded form of the native conformation (20) or a collapsed form of the unfolded state (21). A molten globule-like state is often adopted after an unfolded protein is placed under refolding conditions (22,23), where it may be the preferred state of the unfolded protein (Section 5.2).

4.2 The thermodynamics of unfolding

The thermodynamics of protein unfolding are central to understanding protein stability. They are determined directly by calorimetry, which measures how much energy is required to raise the temperature of a solution of protein, compared to when the protein is absent. This is the *partial heat capacity*, C_p, of the protein in solution. There is a large uptake of heat when a native, folded protein unfolds, but on either side of the unfolding transition the value of C_p remains relatively constant. The unfolded state has the higher heat capacity, and the difference, ΔC_p, is the heat capacity of unfolding. Although the thermodynamic properties of proteins vary greatly, those of a representative set of small proteins demonstrate a number of correlations so that it is possible to give the following general description of thermodynamic parameters for an average protein.

The values of C_p for the folded and unfolded states of a protein are proportional to the non-polar surface area that is exposed to water (24). The explanation for this relationship is that the non-polar groups are hydrated by 'cages' of hydrogen-bonded water molecules (see Chapter 1, Section 2.4), which are melted out and absorb heat as the temperature is increased.

The calorimetric data also give the enthalpy, entropy, and free energy differences between the folded and unfolded states as a function of temperature. Such data have confirmed the two-state nature of unfolding transitions, by demonstrating that the enthalpy change measured calorimetrically is virtually the same as that measured by the temperature dependence of $K_{eq,fold}$.

The stability of a protein varies with temperature (*Figure 3.4*). There is an optimum temperature, found generally to be between −10°C and 35°C, either side of which the stability of the protein decreases. In some cases, unfolding of a protein can be demonstrated simply by lowering the temperature (25). Protein stability is believed to decrease at low temperature because the increased water ordering around the non-polar groups of the unfolded protein favours their solvation (Chapter 1, Section 2.4). This effect is most significant for proteins that have the highest values of ΔC_p, that is, proteins with the highest content of non-polar residues buried in the interior (*Figure 3.4*). At low temperatures, the greater the value of ΔC_p, the less stable the protein (26,27). The thermodynamics of protein unfolding show many similarities to the transfer of hydrophobic molecules from non-polar environments to water (Chapter 1, Section 2.4), but with additional enthalpic and entropic terms that are believed to reflect, respectively, the additional stabilizing interactions of the folded state and the greater conformational entropy of the unfolded state.

4.3 Rationalizing protein stability

The net stability of the folded state of a protein depends upon a complex balance between the many diverse interactions present in the folded state, the much greater conformational disorder of the unfolded state, and the interactions of both states with the solvent. These various factors tend to compensate each other, so the net balance is a small difference between individually large contributions, making it difficult to account fully for the stabilities of folded proteins. This topic is being actively studied by determining the effects on stability of altering the primary structure of polypeptides. Mutagenesis studies clearly demonstrate that the packing of non-polar groups in the interior is important (28–30), whereas

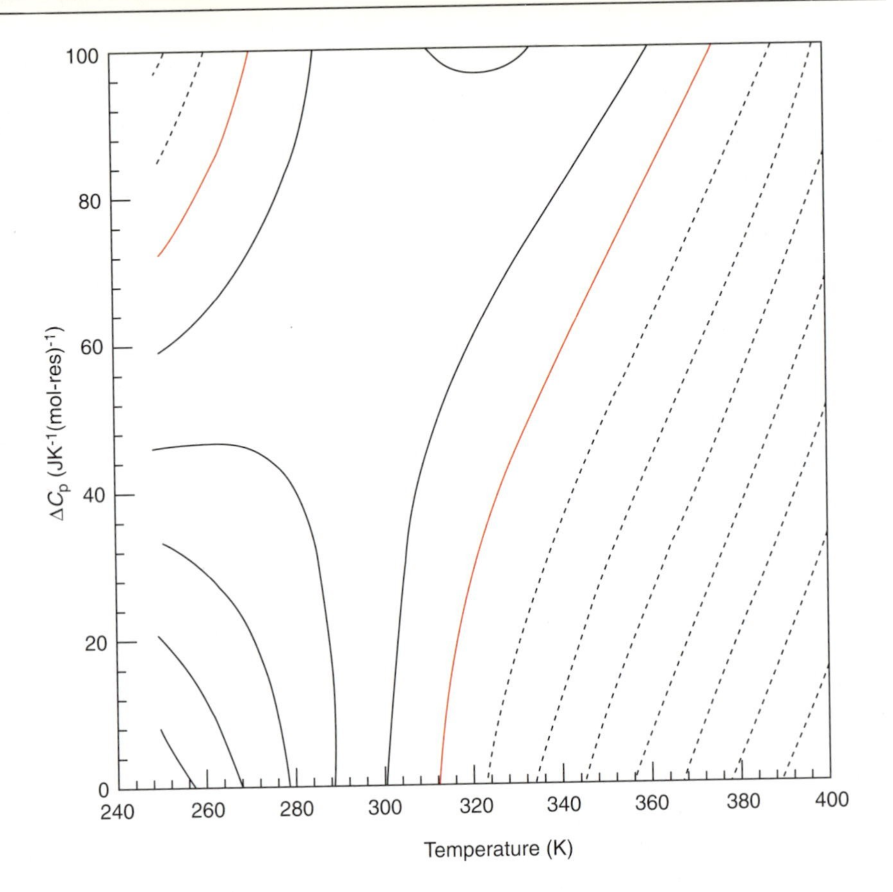

Figure 3.4. The calculated dependence of ΔG_{fold} for a typical protein as a function of its value of ΔC_p and the temperature. The value of ΔC_p is proportional to the amount of non-polar surface area buried in the folded conformation and exposed to water upon unfolding. The orange contour represents $\Delta G_{\text{fold}} = 0$, which is the midpoint of the thermal unfolding transition of a typical protein with the indicated values of ΔC_p. The contour lines represent values of ΔG_{fold} spaced 200 J mol^{-1} residue^{-1} apart, with solid lines representing negative values and the dotted lines representing positive values. The more negative the value of ΔG_{fold}, the greater the protein's stability. Both the ΔG_{fold} and the ΔC_p values are expressed on a per residue basis. Reproduced with permission from ref. 27.

electrostatic interactions play only a very minor role, except when specific, very stable, salt bridges are present (31).

The thermodynamics of protein unfolding (Section 4.2) demonstrate that the burial of non-polar groups within the protein interior is an important contribution to stability. Mutagenesis studies also demonstrate frequently a correlation of stability with hydrophobicity of the buried side-chains (28–30). A number of exceptions to this, however, indicate that the most important factor is not the hydrophobicity of the altered side-chain, but its detailed packing in the interior, which determines the strengths of the van der Waals interactions (32–34). There are also indications that the protein core is not optimally packed to maximize these interactions, in spite of its generally close packing, probably because of the constraints of the covalent and secondary structures.

The contribution of interactions between polar groups, especially hydrogen bonds, to protein stability has generally been disregarded, because of the assumption that their energy in the folded protein is the same as that between water and polar groups in the unfolded protein. This incorrectly ignores the intramolecular nature of the interactions within the folded state, assumes that they operate independently of each other, and disregards any contribution that co-operativity of the interactions may make to their overall strengths. In fact, consideration of the thermodynamics of model systems indicates that hydrogen bonds may play a greater role than non-polar interactions (27). For example, the greater the fraction of non-polar surface buried, the lower is the stability of the protein (26), suggesting that the alternative polar surfaces buried, in hydrogen-bonded form, contribute more to stability (35). For this to be the case, such intramolecular hydrogen bonds would have to be more stable than those formed between the solvent and the unfolded protein; this can only occur if there is co-operativity between them.

To illustrate how co-operativity probably operates (36), consider the interaction of two groups, A and B, in a polypeptide chain (*Figure 3.5*). The single interaction is weak because it decreases the flexibility of the polypeptide chain, and the effective concentration of the groups A and B is low. If it increases the tendency of two other groups, C and D, to come into proximity, this will favour the formation of an interaction between them because their effective concentration with respect to each other is increased by the A–B interaction. Of course, the stabilizing effect is reciprocal, and the interaction C–D must stabilize the interaction A–B to the same extent. This is a solely entropic effect, but there may also be an effect on the enthalpy. For example, hydrogen bonds between groups within a folded protein are usually present essentially all the time, whereas those between the surface of a protein and water are present only part of the time, so the former should have the more favourable enthalpy. The significance of co-operative interactions in maintaining protein structure is demonstrated by the striking two-state nature of the unfolding process (Section 4.1).

To summarize, the present view of protein stability is that it results from both polar and non-polar interactions. Although the individual interactions are very weak, they occur in large numbers in a protein and act together in a co-operative manner. Non-polar interactions, although significant in stabilizing protein structure, probably do not contribute as much to stability as hydrogen bonding.

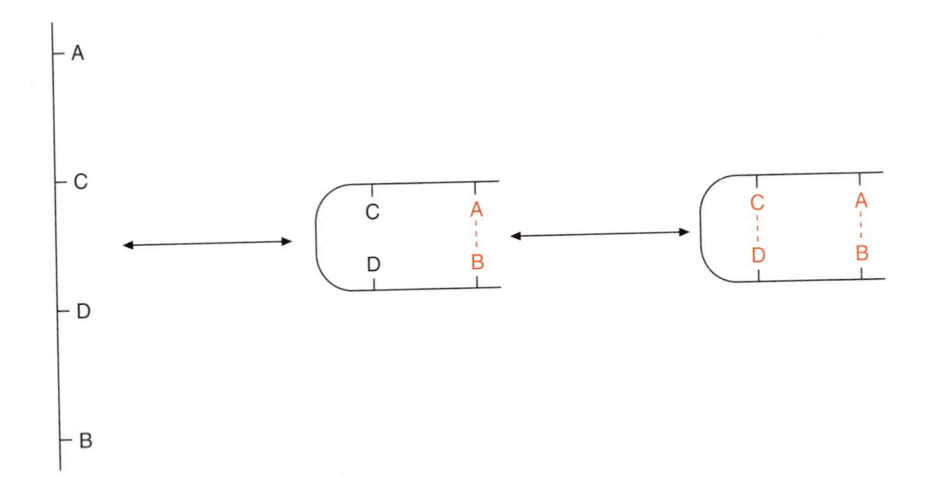

Figure 3.5. Schematic illustration of the principle of entropic co-operativity between interactions. The presence of the interaction between groups A and B brings groups C and D into proximity and favours their interaction. Consequently, the stability of the interaction between groups C and D is greater in the presence of the A····B interaction than in its absence. Exactly the same effect must occur with the converse situation, in which C····D is formed first, as they are linked functions (see *Figure 4.8*). Such co-operativity decreases the stabilities of the intermediate states with just one interaction, and those with either none or all of them tend to predominate at equilibrium.

4.4 Other factors that affect protein stability

A wide range of stabilities are possible with the same three-dimensional fold. For example, proteins from thermophilic organisms, which usually remain folded at temperatures of up to 80 °C, have essentially the same structure as their more normal (mesophilic) counterparts. Given just the structures of two such proteins, it is generally not possible to guess which would have the greater stability, although thermophilic proteins are believed to have more salt bridges. On the other hand, there are instances where special factors are involved in maintaining the stability of the folded state. Most common is the incorporation of a metal ion, or other ligand, into the three-dimensional structure; so long as the ligand does not bind to the unfolded state, the stability of the folded state is directly proportional to the concentration of free ligand, even at high concentrations of ligand when the protein is fully saturated. Other striking examples are proteins from halophilic organisms, which function at high salt concentrations, and often require these conditions for stability. Their conformational stabilities appear to result from the incorporation of ions into their structures (37).

Stabilization provided by disulphide bonds is particularly important in maintaining the folded conformations of some small proteins (38). When reversibly reduced to thiols in the unfolding transition, disulphide bonds can be considered not to be fundamentally different from any other stabilizing interaction, although the bonds are of variable strength, depending upon the reduction/oxidation potential of the environment. When disulphide bonds are retained in the unfolded state, they covalently cross-link the unfolded peptide chain and reduce its conformational entropy. This destabilizes the unfolded state, thereby indirectly further stabilizing the native state.

5. Protein folding

5.1 General characteristics of protein folding

Most small proteins and some larger ones can be refolded from the denatured state, which indicates that their primary sequences contain all the information necessary to determine the final folded structure. However, a small number of proteins such as subtilisin, require the presence of their pro-sequences to fold properly, and the mature processed forms do not readily unfold or refold, unless the pro-sequence is added (39). The pro-sequence thus makes the folded state kinetically accessible, demonstrating the importance of kinetic as well as thermodynamic factors in protein folding.

Small proteins often refold under optimal conditions within seconds or minutes, so it seems unlikely that refolding occurs by the protein randomly passing through all of the very many possible conformational states (Chapter 1, Section 5.1.2). Instead, folding is likely to occur by the protein adopting a limited number of energetically-favourable, non-random intermediate conformations.

The kinetics of the folding process are usually complicated by *cis–trans* isomerization of peptide bonds preceding proline residues, which is intrinsically slow (Chapter 1, Section 5.1.1). In the folded state, individual proline residues can be preceded by either a *cis* or a *trans* peptide bond; with few exceptions (40), the same isomer is generally found at each position in all the folded molecules. However, prolyl peptide bonds isomerize to an equilibrium mixture of *cis* and *trans* forms when the protein unfolds, so that only a fraction of the unfolded molecules have all such proline peptide bond isomers fully compatible with the folded state. These particular unfolded molecules often refold rapidly, whereas refolding of the others is delayed by the required isomerization to the correct isomer. The situation is often complex, however, for some molecules can refold with an incorrect isomer, which subsequently isomerizes, and non-random conformation in the protein can either increase or decrease the rate of isomerization of the peptide bonds.

Except for *cis–trans* isomerization, refolding is generally found to occur with a single rate constant (41). This suggests that there is a single rate-limiting step in the folding pathway that is identical for each protein molecule. More complex kinetics would be expected if different protein molecules folded by different mechanisms, as each would be characterized by a unique rate constant. A common rate-limiting step appears to be followed by all molecules, in spite of the undoubted conformational heterogeneity of the fully unfolded state (Chapter 1,

Section 5.1.2). This is possible because the molecules equilibrate rapidly when placed under refolding conditions between a relatively small number of conformations. This transient unfolded state is actively being characterized, but it frequently appears to be similar to the molten globule state (22,23) (Section 4.1).

The rate-limiting step in folding is that with the overall highest free energy barrier, and involves the overall *transition state* in the folding pathway. In general, the transition state occurs late in the folding process and seems to be a high-energy, distorted form of the fully folded state (41). After this rate-limiting step, only minor conformational adjustments occur. The transition state is being 'mapped' by site-directed mutagenesis to determine which interactions present in the folded state are also present in the transition state (42).

Folding of multi-domain proteins is correspondingly more complex (43). Isolated domains often fold independently to a stable structure. Nevertheless, when two or more are linked in the complete polypeptide, they often interfere with each other's refolding. The final step in such a folding process is the association of the individual domains. It is often the slowest step, so intermediates accumulate that have one or more domains folded but unpaired; they often recognize their complementary partner in another polypeptide chain, so aggregation tends to occur. In contrast, where the stability of a structural domain is highly dependent upon interactions with other domains, the folding process almost assumes the properties of a single co-operative process.

5.2 *Identifying folding intermediates*

Folding intermediates are difficult to identify and to characterize in detail because they are intrinsically unstable and accumulate, at best, only transiently. One possible exception is the molten globule state, which is stable for certain proteins under special conditions (Section 4.1) and which has been shown to be a kinetic intermediate in the folding of a number of proteins (22,23). However, the most detailed information on the nature of folding intermediates comes from proteins whose folding is linked to disulphide bond formation, particularly bovine pancreatic trypsin inhibitor (BPTI). In contrast to other interactions that stabilize the folded conformation, disulphide bonds can be trapped in a stable form (38).

The most pertinent property of BPTI is that it loses its folded conformation when its three disulphide bonds are reduced. It can be refolded at a controlled rate by reforming these disulphide bonds, a process that can be stopped at any time by covalently blocking all the free thiol groups. Analysis of the species with different disulphide bonds that are present during refolding of BPTI as a function of time shows that refolding occurs along a relatively defined pathway, passing through a limited number of disulphide bond intermediates (*Figure 3.6*).

Because reduced BPTI is a very unfolded molecule, all possible disulphide bonds are formed initially, but these rapidly rearrange intramolecularly. Referring to the intermediates by the numbers of the cysteine residues paired in disulphide bonds, intermediate form [30–51] predominates, because it is the most stable of the one disulphide bond intermediates, which are all in rapid equilibrium with each other. It has part of the native conformation, in particular the major

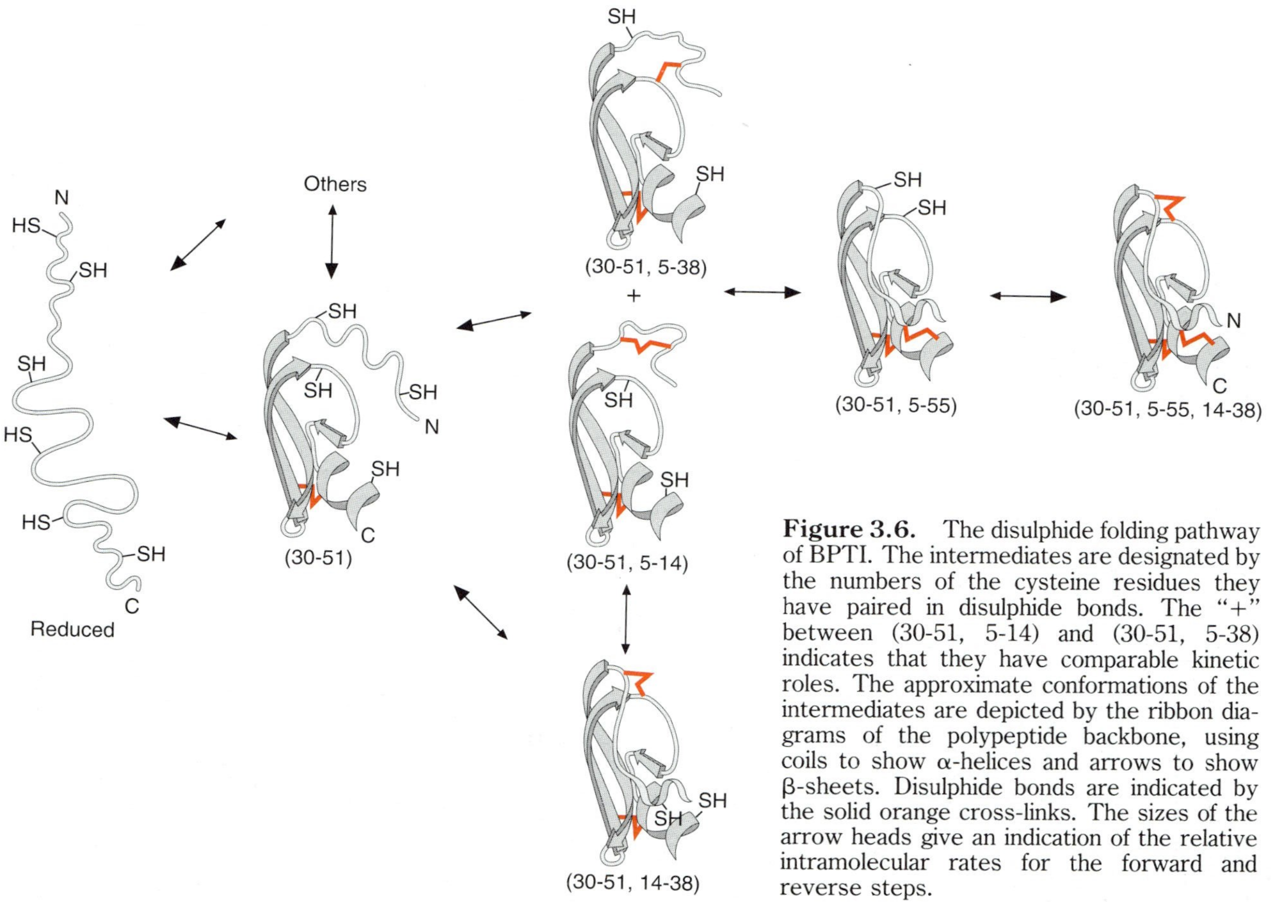

Figure 3.6. The disulphide folding pathway of BPTI. The intermediates are designated by the numbers of the cysteine residues they have paired in disulphide bonds. The "+" between (30-51, 5-14) and (30-51, 5-38) indicates that they have comparable kinetic roles. The approximate conformations of the intermediates are depicted by the ribbon diagrams of the polypeptide backbone, using coils to show α-helices and arrows to show β-sheets. Disulphide bonds are indicated by the solid orange cross-links. The sizes of the arrow heads give an indication of the relative intramolecular rates for the forward and reverse steps.

elements of secondary structure, an α-helix and a β-sheet, which interact to stabilize each other via numerous non-polar interactions, in addition to the disulphide bond between cysteine-30 and cysteine-51.

The most energetically favourable rate-limiting step in BPTI folding is formation of the intermediate [30–51,5–55] via non-native second disulphide bonds. This arises because neither of the intermediate forms [30–51] and [30–51,14–38] readily form the 5–55 disulphide bond, probably because the transition state for forming the 5–55 disulphide bond (which will be buried in the native conformation) has too high an energy. Formation of the 14–38 disulphide bond in the native-like conformation of the [30–51,5–55] intermediate proceeds rapidly, with only very minor conformational adjustments, because the participating thiol groups are held on the surface of the molecule and in the correct orientation with respect to each other.

5.3 Biosynthetic folding

Studying protein folding *in vivo* is even more difficult than *in vitro*, so key questions concerning the temporal relationship between polypeptide biosynthesis on the ribosome and folding have yet to be answered. Individual domains in multi-domain proteins seem to fold while the synthesis of the remainder of the polypeptide is being completed. This may overcome some of the slowness of refolding observed with these proteins *in vitro*, where domains fold simultaneously (Section 5.1). It is not clear, however, whether a small protein or an individual domain starts to fold before its synthesis is complete, for *in vitro* models of the nascent chain, fragments of domains lacking a few C-terminal residues, are generally unfolded (44,45).

A number of proteins are known to catalyse or otherwise assist the folding process. Two slow events, disulphide bond rearrangement (Section 5.2) and *cis–trans* isomerization of peptide bonds preceding proline residues (Chapter 1, Section 5.1.1), are known to be enzyme-catalysed. Disulphide bond rearrangements, and possibly also their formation, are accelerated by an enzyme in the endoplasmic reticulum, *protein disulphide isomerase* (PDI) (46). It catalyses the interchange of the disulphide bonds in a refolding protein, probably using cysteine residues at its active site, until the most stable disulphide-bonded state of the protein is reached. PDI is often found complexed with other enzymes involved in post-translational modifications, most notably prolyl-4-hydroxylase, which is involved in collagen biosynthesis, suggesting that there may be co-ordination between these different events.

The *peptidyl prolyl isomerases* (PPIase) catalyse *cis–trans* isomerization of proline peptide bonds (47). They are ubiquitous proteins that are found in numerous subcellular compartments. At least two distinct classes of PPIase are known, which intriguingly bind two classes of immunosuppressive agents. The relationship between their PPIase activity and immunosuppression is not yet known.

Other proteins, generally termed *chaperones*, probably do not catalyse protein folding but assist it by preventing misfolding (48). The necessity for such assistance is in part apparent from *in vitro* folding studies that show that unfolded and partially folded proteins tend to be insoluble and aggregate, while the subunits of oligomeric proteins often make incorrect interactions. The levels of chaperones are elevated in cells subject to stress, in particular heat shock, possibly because of the increased presence of unfolded or misfolded proteins that must be refolded or targeted for degradation.

One class of chaperone, which includes proteins known as BiP and heat-shock protein (hsp) 70, binds to newly-synthesized proteins and short linear peptides. How the chaperone recognizes the nascent chain is unclear, although it may simply be because the non-polar groups that are normally buried within the folded protein are exposed (49). The bound proteins are rapidly released from the complex in an ATP-dependent reaction and may be passed on to the hsp60-type chaperone.

The hsp60-proteins constitute a second class of chaperones, sometimes referred to as *chaperonins,* which include GroEL of *Escherichia coli* and the hsp60 proteins of mitochondria and chloroplasts. *In vitro* studies of the refolding of a number of proteins that do not refold by themselves have shown that they can be refolded to their native conformations with essentially 100% yield in the presence of stoichiometric amounts of GroEL. GroEL binds to partially folded peptide chains, perhaps in the molten globule state, preventing their aggregation and further folding. Release of the chains, so that they can resume folding, is an ATP-dependent process. Proteins are bound again if they have not completed refolding. The most efficient refolding occurs in the presence of a second, smaller protein, GroES, which normally exists as a ring of seven identical monomers. The GroEL chaperonins are asymmetric double rings of seven monomers each, and GroEL and GroES together are known as GroE. Together, they appear primarily to sequester a single unfolded protein molecule at their centre so that it is unable to interact with, and precipitate, other molecules (50).

Other chaperones assist in the translocation of proteins through membranes (51). This is not required with polypeptides with signal peptides, where the nascent chains are translocated into the endoplasmic reticulum co-translationally. For proteins directed to other organelles or membranes, however, the completed polypeptide chain is released into the cytosol and must be prevented from folding before being translocated. This generally occurs by the unfolded protein interacting with a chaperone, such as an hsp70 protein or the *E. coli* secB protein.

The study of factors involved in protein folding and assembly *in vivo* is one of the most active areas of research, so this area is likely to see rapid advances in the future.

6. Further reading

Membrane proteins

Israelachvili, J. N. (1980) Physical principles of membrane organization. *Q. Rev. Biophys.*, **13**, 121.

von Heijne,G. (1988) Transcending the impenetrable: how proteins come to terms with membranes. *Biochim. Biophys. Acta,* **947**, 307.

Jennings,M.L.(1989) Topography of membrane proteins. *Annu. Rev. Biochem.,* **58**, 999.

Protein stability

Schellman,J.A. (1987) The thermodynamic stability of proteins. *Annu. Rev. Biophys. Biophys. Chem.,* **16**, 115.

Timasheff,S.N. and Arakawa,T. Stabilization of protein structure by solvents. In Creighton,T.E. (ed.). *Protein Structure: a Practical Approach.* IRL Press, Oxford, p.331.

Privalov,P.L. (1979) Stability of proteins. Small globular proteins. *Adv. Protein Chem.,* **33**, 167.

Privalov,P.L. (1982) Stability of proteins. Proteins which do not present a single cooperative system. *Adv. Protein Chem.,* **35**, 1.

Privalov,P.L. (1989) Thermodynamic problems of protein structure. *Annu. Rev. Biophys. Biophys. Chem.,* **18**, 47.

Protein folding

Goldenberg, D.P. (1988) Genetic studies of protein stability and mechanisms of folding. *Annu. Rev. Biophys. Biophys. Chem.,* **17**, 481.

Creighton,T.E. (1990) Protein folding. *Biochem. J.,* **270**, 1.

Kim,P.S. and Baldwin,R.L. (1990) Intermediates in the folding reactions of small proteins. *Annu. Rev. Biochem.,* **59**, 631.

7. References

1. Liljas,L. (1986) *Prog. Biophys. Mol. Biol.,* **48**, 1.
2. Rupley,J.A. and Careri,G. (1991) *Adv. Protein Chem.,* **41**, 37.
3. Otting,G., Liepinsh,E. and Wüthrich,K. (1991) *Science,* **254**, 974.
4. Teeter,M.M. (1984) *Proc. Natl. Acad. Sci. USA,* **81**, 6014.
5. Arakawa,T. and Timasheff,S.N. (1985) *Methods Enzymol.,* **114**, 49.
6. Matthew,J.B., (1985) *Annu. Rev. Biophys. Bioeng.,* **14**, 387.
7. Rodgers,N.K. (1986) *Prog. Biophys. Mol. Biol.,* **48**, 37.
8. Sharp,K.A. and Honig,B. (1990) *Annu. Rev. Biophys. Biophys. Chem.,* **19**, 301.
9. Rees,D.C., DeAntonio,L., and Eisenberg,D. (1989) *Science,* **245**, 510.
10. Engelman,D.M., Steitz,T.A., and Goldman,A. (1986) *Annu. Rev. Biophys. Biophys. Chem.,* **15**, 321.
11. Jennings,M.L. (1989) *Annu. Rev. Biochem.,* **58**, 999.
12. Deisenhofer,J. and Michel,H. (1991) *Annu. Rev. Biophys. Biophys. Chem.,* **20**, 247.
13. Rees,D.C., Komiya,H., Yeates,T.O., Allen,J.P., and Feher,G. (1989) *Annu. Rev. Biochem.,* **58**, 607.
14. Weiss,M.S., Abele,U., Weckesser,J., Welte,W., Schiltz,E., and Schulz,G.E. (1991) *Science,* **254**, 1627.
15. Popot,J.-L. and de Vitry,D. (1990) *Annu. Rev. Biophys. Biophys. Chem.,* **19**, 369.
16. Breslow,R. and Guo,T. (1990) *Proc. Natl. Acad. Sci. USA,* **87**, 167.
17. Schmid,F.X. (1989) In Creighton,T.E. (ed.), *Protein Structure: a practical approach.* IRL Press, Oxford, p.251.
18. Pace,C.N. (1990) *Trends Biochem. Sci.,* **15**, 14.
19. Kuwajima,K. (1989) *Proteins: Struct. Funct. Genet.,* **6**, 87.

20. Baum,J., Dobson,C.M., Evans,P.A., and Hanley,C. (1989) *Biochemistry*, **28**, 7.
21. Ewbank,J.J. and Creighton,T.E. (1991) *Nature*, **350**, 518.
22. Ikeguchi,M., Kuwajima,K., Mitani,M., and Sugai,S. (1986) *Biochemistry*, **25**, 6965.
23. Ptitsyn,O.B., Pain,R.H., Semisotnov,G.V., Zerovnik,E., and Razgulyaev,O.I. (1990), *FEBS Lett.*, **262**, 20.
24. Privalov,P.L. and Makhatadze,G.I. (1990) *J. Mol. Biol.*, **213**, 385.
25. Privalov,P.L. (1990) *CRC Crit. Rev. Biochem.*, **25**, 281.
26. Murphy,K.P., Privalov,P.L., and Gill,S.J. (1990) *Science*, **247**, 559.
27. Murphy,K.P. and Gill,S.J. (1991) *J. Mol. Biol.*, **222**, 699.
28. Matthews,B.W (1987) *Biochemistry*, **26**, 6885.
29. Shortle,D. (1989) *J. Biol. Chem.*, **264**, 5315.
30. Pakula,A.A. and Sauer,R.T. (1989) *Annu. Rev. Genet.*, **23**, 298.
31. Anderson,D.E., Becktel,W.J., and Dahlquist,F.W. (1990) *Biochemistry*, **29**, 2403.
32. Kellis,J.T., Nyberg,K., and Fersht,A.R. (1989) *Biochemistry*, **28**, 4914.
33. Sandberg,W.S. and Terwilliger,T.C. (1989) *Science*, **245**, 54.
34. Lim,W.A. and Sauer,R.T. (1991) *J. Mol. Biol.*, **219**, 359.
35. Creighton,T.E. (1990) *Curr. Opinion Struct. Biol.*, **1**, 5.
36. Creighton,T.E. (1983) *Biopolymers*, **22**, 49.
37. Zaccai,G. and Eisenberg,H. (1990) *Trends Biochem. Sci.*, **15**, 333.
38. Creighton,T.E. (1990) *Biochem. J.*, **270**, 1.
39. Baker,D., Sohl,J.L., and Agard,D.A. (1992) *Nature*, **356**, 263.
40. Chazin,W.J., Kördel,J., Drakenberg,T., Thulin,E., Brodin,P., Grundström,T., and Forsén,S. (1989) *Proc. Natl. Acad. Sci. USA*, **86**, 2195.
41. Creighton,T.E. (1988) *Proc. Natl. Acad. Sci. USA*, **85**, 5082.
42. Matouschek,A., Kellis,J.T., Serrano,L., and Fersht,A.R. (1989) *Nature*, **340**, 122.
43. Jaenicke,R. (1987) *Prog. Biophys. Mol. Biol.*, **49**, 117.
44. Taniuchi,H. (1970) *J. Biol. Chem.*, **245**, 5459.
45. Shortle,D. and Meeker,A.K. (1989) *Biochemistry*, **28**, 936.
46. Freedman,R.B. (1989) *Cell*, **57**, 1069.
47. Fisher,G. and Schmid,F.X. (1990) *Biochemistry*, **29**, 2205.
48. Gething,M.-J. and Sambrook,J. (1992) *Nature*, **355**, 33.
49. Flynn,G.C., Pohl,J., Flocco,M.T., and Rothman,J.E. (1991) *Nature*, **353**, 726.
50. Martin,J., Langer,T., Boteva,R., Schramel,A., Horwich,A.L., and Hartl,F.-U. (1991) *Nature*, **352**, 36.
51. Ellis, R.J. (ed.) (1990) *Seminars Cell Biol.*, **1**,1.

4

Ligand binding and protein function

1. Introduction

Proteins must be able to recognize and bind other molecules specifically in order to fulfil virtually all of their functional roles, for example as enzymes, receptors, or hormones. The 'target' molecules will be classed together here as *ligands*, even though they may vary in size and complexity from an electron to another protein or other macromolecule. This chapter examines the general principles of how molecular recognition is achieved and illustrates them by considering some specific examples. It is especially important to identify the determinants of ligand binding, as the functions of many newly identified proteins can frequently be deduced just from their amino acid sequence if residues that make up binding sites for specific ligands can be recognized in that sequence.

2. Ligand binding

2.1 Studying protein–ligand interactions

The basic parameter that characterizes the affinity of a protein for a ligand is the overall equilibrium constant for binding, expressed as either the *association constant* (K_a) or its reciprocal, the *dissociation constant* (K_d). The latter has the advantage of representing the free ligand concentration at which the protein binding site is half saturated. The association and dissociation constants are often used to calculate the free energy of binding and can be expressed in the form:

$$\Delta G_{\text{bind}} = -\text{R}T \log_e(K_a) = \text{R}T \log_e(K_d) \qquad (4.1)$$

The value of ΔG_{bind} defined in this way depends upon the concentration units used and is pertinent only when the ligand is present at one unit of that concentration. For example, if the units used for K_a and K_d are mol/l, the value of ΔG_{bind} calculated will only apply when the concentration of the free ligand is one molar. The significance of ΔG_{bind} is further reduced when it is realized that it is not simply the sum of the free energy contributions of all the interacting groups, because binding involves other factors, especially the loss of entropy of the ligand and protein upon binding (1).

The specificity of binding is determined by the relative binding constants of variant ligands. Varying the covalent structure of the ligand or altering the protein by mutagenesis, gives some indication of how particular chemical groups contribute to the strength of binding. For example, comparison of the affinities of a ligand and a protein when group A is present and absent in either of them, K_a^{+A} and K_a^{-A}, respectively, is frequently used (2) to measure the free energy contribution of group A, $\Delta\Delta G_{\text{bind,A}}$, to the interaction:

$$\Delta\Delta G_{\text{bind,A}} = -RT \log_e(K_a^{+A}/K_a^{-A}) = -RT \log_e(K_d^{-A}/K_d^{+A}) \qquad (4.2)$$

Although such measurements are very useful, they may be misleading because the changes may alter the mode of binding of the ligand, particularly if it is the ligand that is altered.

To identify the binding site of a protein of unknown structure, chemically reactive groups may be incorporated into the ligand to react preferentially with the residues present in the binding site, so that they can subsequently be identified by protein chemical methods (3). With enzymes, where the ligand is transformed chemically, there is also the possibility of using *suicide substrates*, which produce a reactive intermediate that reacts with nearby groups in the protein. Large ligands, especially proteins, may be cross-linked to the binding site by using bi-functional reagents that have chemical groups that can react with the protein and with the ligand at each end, separated by a spacer of appropriate length (4).

The most detailed information on the nature of the interaction between protein and ligands, however, comes from X-ray diffraction or NMR studies on the protein–ligand complex. The complex can be crystallized, or a small ligand can be diffused into a crystal of just the protein. If the crystal structure of the protein is known, the ligand introduced can be located by the *difference Fourier* method, which assumes the phases of the reflections to be unchanged and uses only the changes in the amplitudes. In NMR studies, the ligand and the protein can be differentiated by incorporating different isotopic labels into them.

2.2 General properties of protein–ligand interactions

Individual polypeptide chains generally have a single binding site for a particular ligand, although multiple sites are likely to be present if the chain has arisen by gene duplication, and there are many other exceptions to this general rule. With polypeptides that bind two or more different ligands, each is often bound to a separate domain of the protein, but there are also many instances where individual domains can bind up to three different ligands simultaneously.

The interactions between a protein and ligand are usually similar to those involved in stabilizing the folded states of proteins (Chapter 1, Section 2 and Chapter 3, Section 4.3) and are equally difficult to rationalize. The relative affinities with which different ligands bind to a protein are presumed to be proportional to their steric and chemical complementarity to the binding site (5–7). The binding sites on proteins seem designed to maximize the strengths and specificities of the interactions with the ligand, while still permitting it to associate and to dissociate. For example, the smallest ligands and prosthetic groups often bind within the interior of the protein, where they are surrounded by protein groups; such ligands are assumed to be able to get into and out of internal sites as a result of their small size and fluctuations of the protein structure. The binding sites for larger ligands are generally depressions or clefts on the protein's surface. Where the ligand is larger than the protein, for example a molecule of DNA, the area of the interaction may be maximized by the protein binding into a cleft in the ligand.

Binding sites of proteins do not generally undergo substantial alteration upon binding a ligand, although some adjustment of the overall structure may be required to facilitate the process of binding and dissociation. The analogy is often made that ligand binding is like a key (the ligand) fitting into a lock (the protein). When the ligand is flexible before binding, it must adapt its three-dimensional structure to fit into a binding site. Much more rarely does the protein adapt its structure substantially to bind the ligand. All conformational changes that are necessary in either the ligand or the protein diminish the observed binding affinity. Perhaps for this reason, conformational changes occur in proteins only when they are of functional importance, as in the allosteric regulation of ligand binding (Section 8.1) or in *induced fit,* which ensures that chemical reactions occur on enzymes only when all of the appropriate substrates are present (8). These conformational changes in proteins are generally limited to quaternary structure rearrangements of domains or subunits relative to each other (9).

The specificity of ligand binding is an essential parameter governing the biological activities of proteins. In some instances, a low specificity of interaction is an advantage, for example, to allow an enzyme to have a broad substrate specificity. At the other extreme, stringent specificity in processes such as the biosynthesis of a protein or DNA is necessary to prevent the misincorporation of incorrect amino acids or nucleotides. In the latter cases, the ability of the participating enzymes to discriminate between the individual amino acids or nucleotides is insufficient to produce an acceptable error rate, and special 'proof-reading' mechanisms are employed to discriminate against incorrect ligands (10).

3. The relationship between protein structure and ligand binding

Homologous proteins, with very similar three-dimensional structures, often have similar ligand binding properties, but this cannot be assumed always to be the case. The specificity with which a ligand binds to a protein is usually determined by a small number of residues at or near the binding site, so variation of just a few residues therefore allows homologous proteins to have very different ligand binding specificities. The most extreme example of this occurs in the immunoglobulins whose specificity relies on the amino acid sequences of just six surface loops (11). Many millions of antibodies with different specificities are generated by varying the sequences of these six loops, yet the overall structures of all immunoglobulins are very similar (12, 13). In another example, the protein haptoglobin is homologous to the serine proteases, but is not a protease. Instead, its function is to bind haemoglobin dimers in the process of haemoglobin degradation (14).

To identify the function of a new protein from the homology of its amino acid sequence to that of a known protein, it is necessary to determine that the residues involved directly in binding a certain ligand are conserved. For example, many proteins that bind calcium ions (Section 4), electrons (Section 5), nucleotides (Section 6), and nucleic acids (Section 7) are identifiable from their sequences alone.

The ability of different proteins to bind similar ligands does not mean that they have a related three-dimensional structure, or even that the mechanism of ligand binding is the same in each protein. Globins and the cytochrome *c*-related proteins all bind haem, but they do so within different three-dimensional structures. Indeed, cytochrome c_3 has four haem binding sites, each of which binds the prosthetic group by a different mechanism.

It is more difficult to understand situations in which there are similarities between two proteins in their mechanism of binding ligands, but little evidence that they had a common evolutionary origin (15). Does this indicate extreme evolutionary divergence, or have the similarities arisen by convergence?

4. Calcium-binding proteins

Proteins bind specifically a variety of metal ions. The groups of the protein involved directly in the binding are usually polar groups that have intrinsic affinities for the particular ion. For example, calcium ions are usually liganded by oxygen molecules of the protein, zinc ions by nitrogen atoms, particularly of histidine side chains, and ferrous and ferric ions by the sulphur atoms of cysteine residues. In all cases, however, a number of them are arranged in space by the protein conformation to create a binding site with dramatically increased affinity and specificity (16, 17).

Calcium ions, in particular, play important roles in cells, for example in signal transmission, which requires that they interact specifically with certain proteins. Some Ca^{2+}-binding proteins, such as calmodulin, function as sensors of alterations in Ca^{2+} levels, while others, such as the parvalbumins and calbindin, appear to buffer the cellular concentration of the free ion. In other proteins, the Ca^{2+} serves no functional role and may simply be involved in stabilizing the folded structure.

A variety of Ca^{2+}-binding sites are observed in proteins, but the *EF hand* Ca^{2+}-binding motif is frequently observed in regulatory proteins (18, 19). It was first identified in parvalbumin, where the E and F helices of this protein participate in the binding site. These two helices are roughly perpendicular to each other and are connected by a short loop of 12 residues. The spatial relationship of these structural elements resembles a right hand with the thumb and first finger stretched out, representing the helices, and a closed third finger representing the loop; hence the term 'EF hand'. Each Ca^{2+} co-ordinates with a pentagonal bipyramidal array of oxygen atoms held in the correct spatial orientation by the protein structure. The oxygen atoms come from the side chains of aspartic acid, asparagine, glutamic acid, threonine, and serine residues at positions 1, 3, 5, 9, and 12 of the loop, from the backbone carbonyl of residue 7, and from a water molecule. The special spatial and sequential arrangements of these residues allow the EF hand to be recognized from just the primary structures of proteins. From two to eight EF hands may be present in a protein; they are usually positioned in pairs, which appears to increase the affinity for Ca^{2+}.

Proteins that function as sensors of Ca^{2+} levels relay alterations through Ca^{2+}-induced changes in their conformations. In calmodulin and troponin C (*Figure 4.1*), two EF hands are located at each end of a long α-helix (19). The binding sites show differences in affinity for Ca^{2+}, with the sites at one end of the helix (I and II in *Figure 4.1*) showing a lower affinity. Ca^{2+} binding at the low affinity sites induces a conformational change in the protein that causes the EF hands at each end of the helix to move relative to each other. A consequence of this in troponin C is that a patch of non-polar residues is exposed on the surface. Proteins whose activities are regulated by Ca^{2+} via calmodulin or troponin C may interact specifically with this non-polar site.

Other Ca^{2+}-binding sites that do not utilize EF hands also co-ordinate the Ca^{2+} ion with multiple oxygen atoms, but they do not otherwise have common structural properties.

5. Redox proteins

Redox proteins bind electrons reversibly. The only natural amino acids capable of doing this are cysteine residues, which effectively release two electrons when making a disulphide bond (Chapter 1, Section 3). Most redox proteins, however, utilize a prosthetic group containing a metal ion that can exist in different redox states, such as Cu^{+}/Cu^{2+} or Fe^{2+}/Fe^{3+}. The affinity of the prosthetic group for electrons is modulated by its interaction with the protein. Thus, haem proteins using the same iron atom and haem group exhibit affinities for electrons that range over 24 orders of magnitude in association constant, or 135 kJ mol^{-1} in binding energy.

One way in which proteins achieve this modulation is by varying the environment of the metal ion, particularly its co-ordination by amino acid side chains (20). Different oxidation states of the metal ion prefer different patterns of co-ordination; for example Cu^{+} prefers tetrahedral co-ordination, whereas Cu^{2+} prefers planar or octahedral co-ordination. Copper ions tetrahedrally co-ordinated

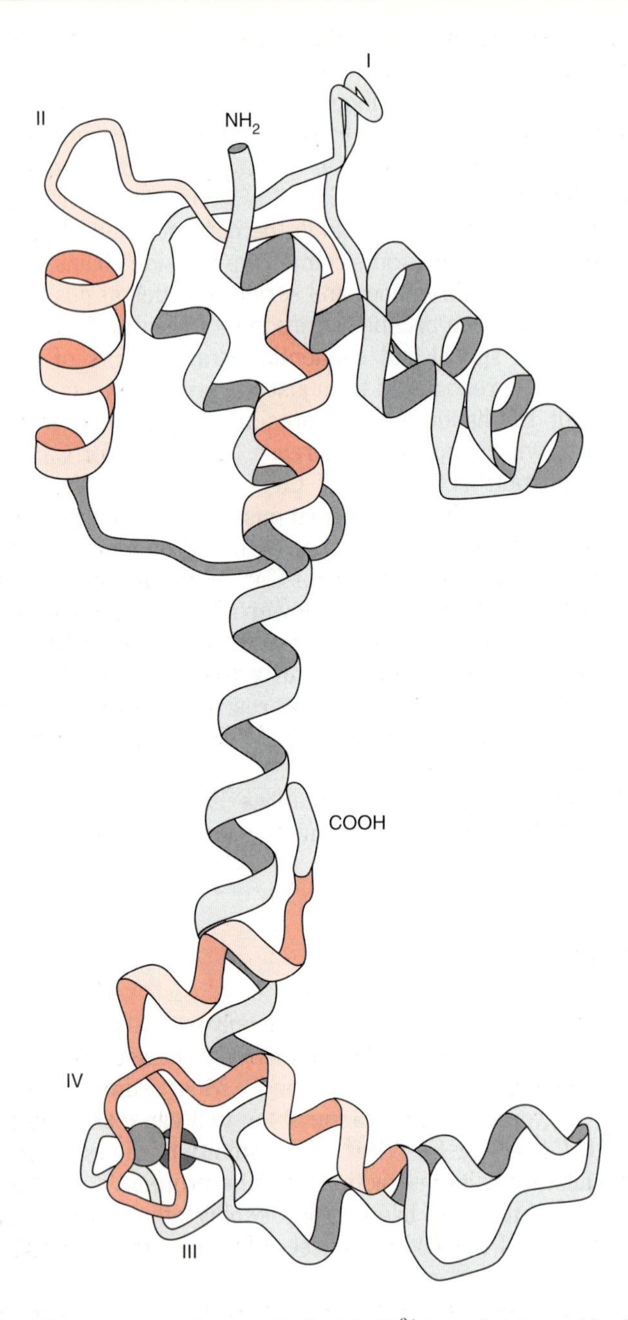

Figure 4.1. The structure of troponin C with Ca^{2+} bound at two of its four EF hands. Sites II and IV have been highlighted in orange. Note the differences in their conformation due to the presence of Ca^{2+} at sites III and IV and its absence at sites I and II. Reproduced with permission from Herzberg, O. and James, M.N.G. (1988) *J. Mol. Biol.*, **203**, 761.

by a protein will prefer to be Cu^+, rather than Cu^{2+}, and so will have a higher affinity for electrons than a copper atom that is planar or octahedrally co-ordinated.

To achieve such control over the redox potential of a prosthetic group, the protein side chains co-ordinating it must be held in place by a protein structure that, overall, is relatively rigid. Otherwise, an unfavourable oxidation state might cause the protein to unfold or to release the prosthetic group because of the structural strain induced. Any alteration induced in the protein structure to accommodate the unfavourable oxidation state more readily would limit the ability of the protein structure to modulate the affinity of the prosthetic group.

6. Nucleotide-binding proteins

A number of methods of binding nucleotides have evolved in proteins, but two related folds have been strongly conserved, one for binding mononucleotides, the other for dinucleotides, such as the co-factors NAD^+ (nicotinamide adenine dinucleotide) and FAD (flavin adenine dinucleotide). The *dinucleotide-binding fold,* also called *Rossmann fold,* was first identified in a number of NAD^+-requiring dehydrogenases, which used different substrates and are structurally similar only in their binding of dinucleotides (21). The dinucleotide fold is based upon two linked β-α-β-α-β units, which form a six-stranded parallel β-sheet (*Figure 4.2*). Each β-α-β-α-β unit binds one of the nucleotide units of the dinucleotide. Different proteins with this fold bind NAD^+ in very similar ways, even though they have very different amino acid sequences within the binding site. There are some similarities in sequence, however, for within each N-terminal β-α-β unit (β_A-α_B-β_B in *Figure 4.2*) there is a highly conserved sequence, Gly-X-Gly-X-X-Gly, which makes it possible to identify nucleotide-binding domains from just the amino acid sequence (22). The first two glycine residues are in the loop between a β-strand and the following α-helix at positions that cannot accommodate an amino acid side chain: they interact with the pyrophosphate group of NAD^+. The third glycine is located in the α-helix, where the absence of a side chain permits especially close packing of the β-α-β unit. There are also six positions that always have hydrophobic residues, because they form the hydrophobic core between the α-helix and the two β-strands. A conserved glutamic acid or aspartic acid residue interacts through a hydrogen bond with the 2′-hydroxyl of the ribose of the nucleotide. The negatively charged phosphates of the nucleotide interact favourably with the positive charge of the helix dipole. The adenine moiety binds in a hydrophobic cleft, while the nicotinamide ring binds so that one face is in a polar environment and the other interacts with non-polar amino acid side chains.

The similar folded structures and mechanisms of interaction with the co-factor of all the dinucleotide-binding proteins suggest that they are evolutionarily related, even though there is little amino acid sequence similarity other than that described for structural and functional reasons.

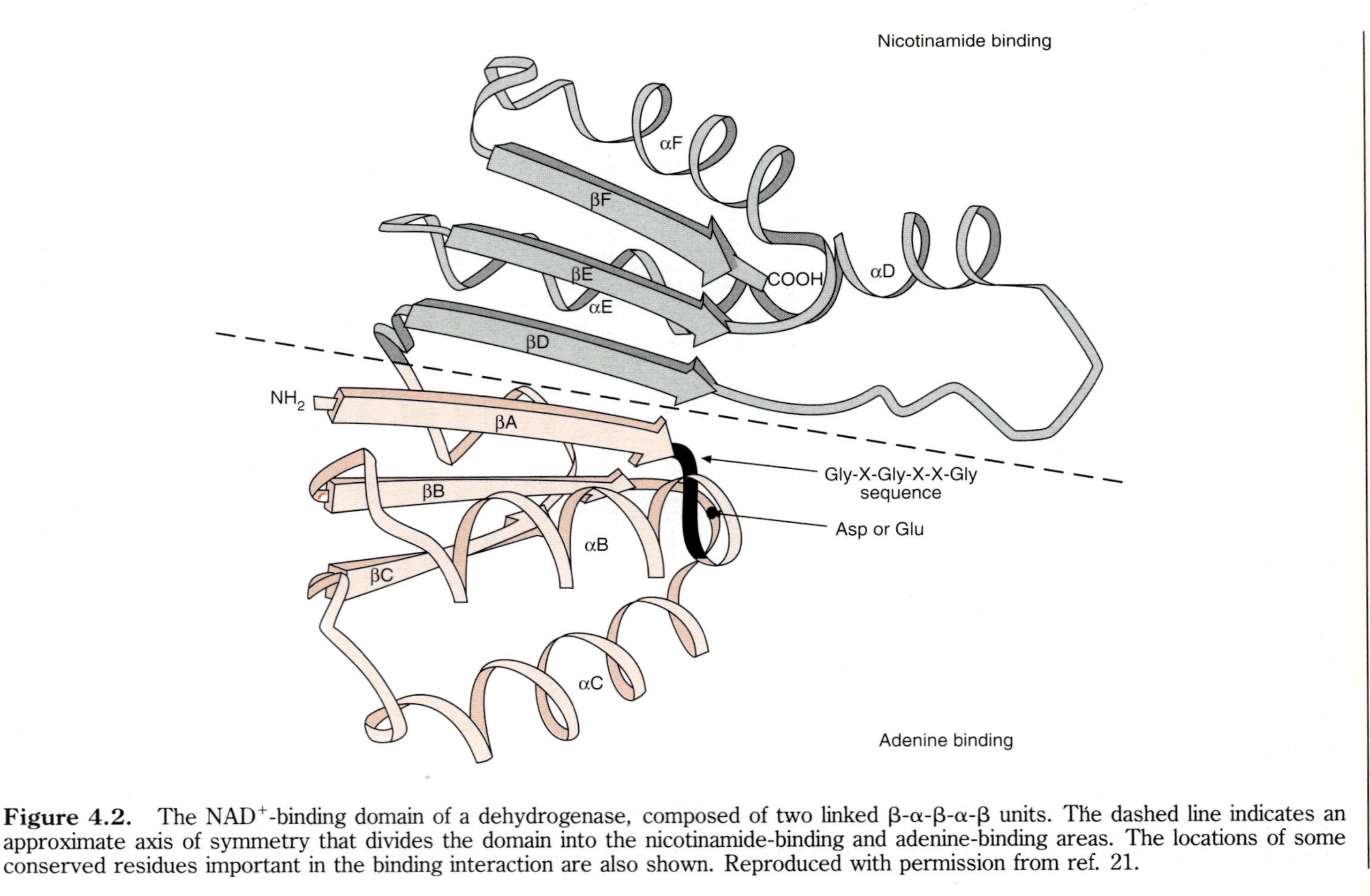

Figure 4.2. The NAD$^+$-binding domain of a dehydrogenase, composed of two linked β-α-β-α-β units. The dashed line indicates an approximate axis of symmetry that divides the domain into the nicotinamide-binding and adenine-binding areas. The locations of some conserved residues important in the binding interaction are also shown. Reproduced with permission from ref. 21.

Minor alterations to the NAD^+-binding site permit binding of $NADP^+$ (23). The third glycine of the consensus sequence becomes an alanine, so that the β-sheet–α-helix packing is no longer so close and the additional, 2′-phosphate of $NADP^+$ can be accommodated. An arginine substitutes for the conserved glutamic acid or aspartic acid residue in the NAD^+-binding site and interacts with the 2′-phosphate group.

A similar β-α-β-α-β fold binds the AMP (adenosine monophosphate) unit of FAD, but there are additional specific interactions in the second β-α-β unit with the ribitol and flavin moieties of the co-factor (24).

The well conserved *mononucleotide-binding fold* was first recognized in adenylate kinase (*Figure 4.3*). It has some similarities to the dinucleotide-binding fold, as it is also a β-α-β-α-β unit and shows a similar sequence at the binding site, Gly-X-X-Gly-X-Gly-Lys, although with a very different structure. In this case, all three glycines occur in a large loop that forms a 'giant anion hole' in which the phosphate of the mononucleotide is accommodated (25). The second and third glycines seem to be particularly important in their positions, which cannot accommodate residues with side chains. Again, the positive helix dipole interacts with the phosphate groups.

A number of other mononucleotide-binding folds have also been found (26). In the protein kinases, the nucleotide phosphates again bind in an anion hole formed by a glycine-rich sequence. In contrast to the adenylate kinase fold, however, this loop links two anti-parallel β-strands. The protein kinase fold is distinguished from that of adenylate kinase by binding the adenosine differently, and by locating the lysine that interacts with the nucleotide in a different part of the secondary structure. The dissimilarity between these mononucleotide folds suggests that any common features of the ligand binding mechanism may have arisen by convergent evolution.

7. DNA-binding proteins

Proteins that bind to DNA are of special importance, for example, in the regulation of gene expression and the control of DNA replication. Some proteins simply bind non-specifically to the sugar–phosphate backbone of DNA through electrostatic interactions. More remarkable are those proteins that recognize very specific DNA nucleotide sequences.

The nucleotide bases of DNA lie at the core of the double helix, and only their edges are exposed to solvent, primarily within the major groove of the double helix. Consequently, the sites of proteins that interact specifically with DNA project away from the protein surface so that they can penetrate the DNA major groove. There they interact with the edges of the nucleotide bases, generally by polar interactions, especially hydrogen bonds (27).

Of crucial importance with DNA-binding proteins is the way in which their affinities for DNA are modulated, because this is how they play a regulatory role.

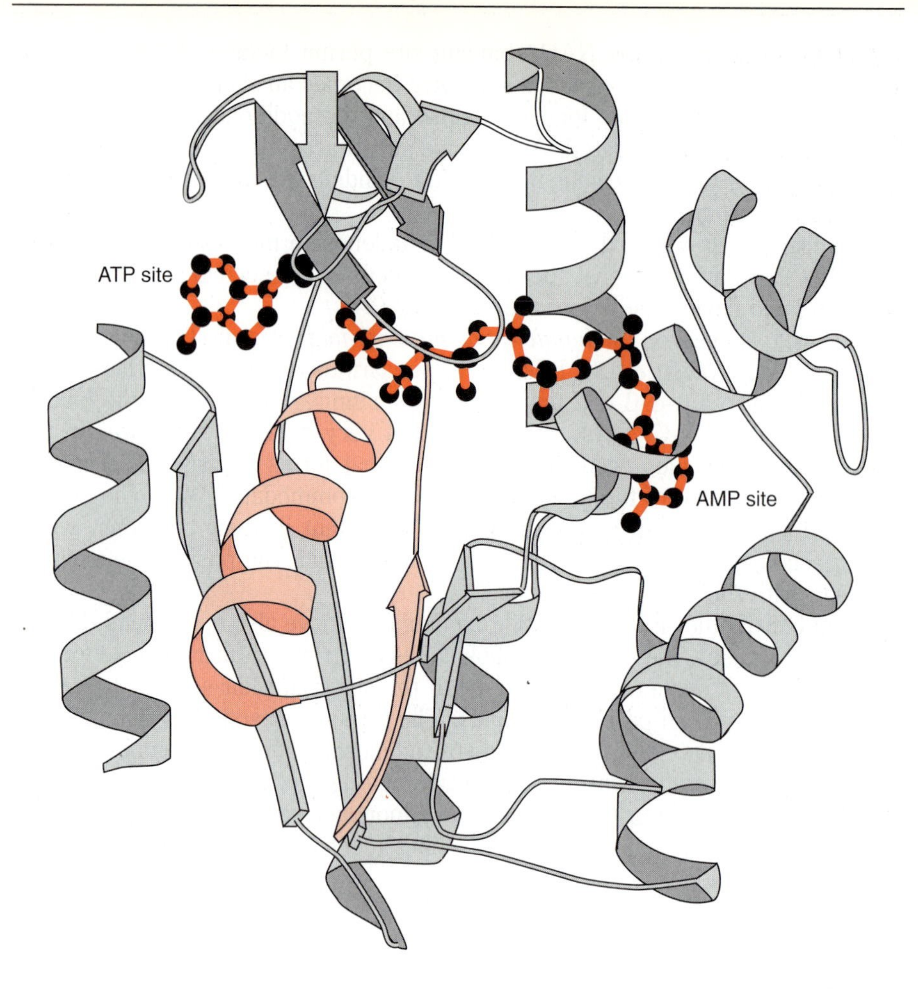

Figure 4.3. The structure of adenylate kinase. Adenylate kinase catalyses the conversion of one ATP and one AMP molecule into two molecules of ADP. Shown in orange is a molecule of P^1, P^5-bis (adenosine-5′) pentaphosphate, a non-hydrolysable analogue of ATP linked to AMP, which was used in the crystallographic analysis of ligand binding to the protein; the ATP-binding site is at the left of the protein and the AMP-binding site at the right. The terminal phosphate of the ATP molecule binds at the 'giant anion hole', which is the loop connecting the β-sheet and α-helix shown in light orange. Reproduced with permission from Muller, C.W. and Schulz, G.E. (1992). *J. Mol. Biol.*, **224**, 159.

Generally, the affinity is greatly enhanced or diminished by the presence of another regulatory ligand. Binding of the regulatory ligand and of the DNA are linked functions (see *Figure 4.8*), so binding of each ligand must affect the affinity of the other to exactly the same extent.

7.1 Helix–turn–helix proteins

The best characterized DNA-binding motif is the helix–turn–helix (*h-t-h*) that is present in proteins from both prokaryotes and eukaryotes (28). This structural unit projects away from the rest of the protein to allow one of the helices, the recognition helix, to bind in the major groove of the DNA (*Figure 4.4a*). The other helix stabilizes the structural motif by packing against the recognition helix, primarily through interactions between non-polar side chains. This is especially important for the structural integrity of the motif, as it can make few other stabilizing interactions with the rest of the protein. For this reason, the residues participating in this packing interaction are highly conserved even in proteins that, apart from the presence of the motif, have little other sequence or structural similarity. The *h-t-h* motif is found embedded in domains of remarkably varied structure (29).

Recognition of the various DNA base pairs is achieved primarily by hydrogen bonds from amino acid side chains, especially those of arginine, asparagine, aspartic acid, glutamine, and glutamic acid residues, which can make multiple hydrogen bonds (*Figure 4.4b*). Water molecules are sometimes involved as bridges in these hydrogen bonds. Van der Waals contacts also participate in the interaction between protein and nucleotides. However, there is no simple code relating the amino acid sequence of the *h-t-h* motif to the sequence of the DNA it recognizes; several side chains can interact with a given base pair, and more than one base pair can be in contact with a given side chain.

Many DNA-binding proteins are dimers that contain two equivalent *h-t-h* motifs separated by 34 Å, the pitch of the DNA helix (*Figure 4.4a*). The two identical *h-t-h* motifs recognize identical or similar DNA sequences, so the binding site on DNA is usually palindromic, with the same nucleotide sequence running in opposite directions on the two DNA strands; each monomer binds to one half of the palindromic sequence. The ways in which the affinities of these proteins are modulated are best known in the case of *trp* repressor protein, which regulates the transcription of the *trp* operon in response to changes in the concentration of the amino acid tryptophan (30). The affinity of the *trp* repressor is controlled by adjustment of the distance between its two *h-t-h* motifs. In the absence of tryptophan, they are separated by 26 Å, so the repressor does not bind to the appropriate segment of DNA and the operon is transcribed. Binding of the amino acid tryptophan to the repressor induces a conformational change in the dimeric structure of the protein that alters the separation of the motifs to 34 Å. This is optimal for their binding to DNA, which turns off transcription of the operon.

A number of reports indicate that the binding of *h-t-h* proteins can alter the local conformation of the DNA structure, causing it to bend or kink (31). This causes the major groove to contract and the minor groove to expand, which can alter the affinities of other nearby DNA–protein interactions.

7.2 Zinc-containing DNA recognition elements

A number of proteins have the common feature that they utilize Zn^{2+} as an integral part of their DNA-binding folds. The first of these to be recognized was the zinc finger motif, which, like the *h-t-h* motif, utilizes an α-helix that protrudes

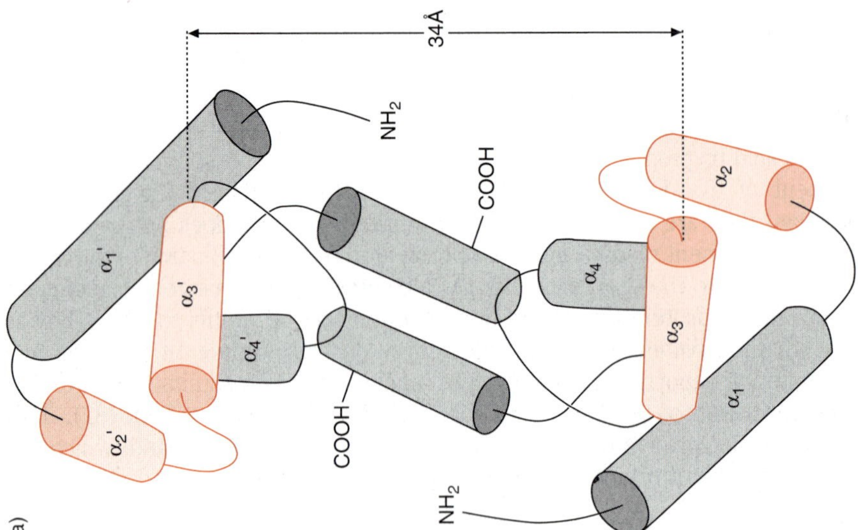
a)
34Å
NH2
COOH
COOH
NH2
α1'
α2'
α3'
α4'
α1
α2
α3
α4

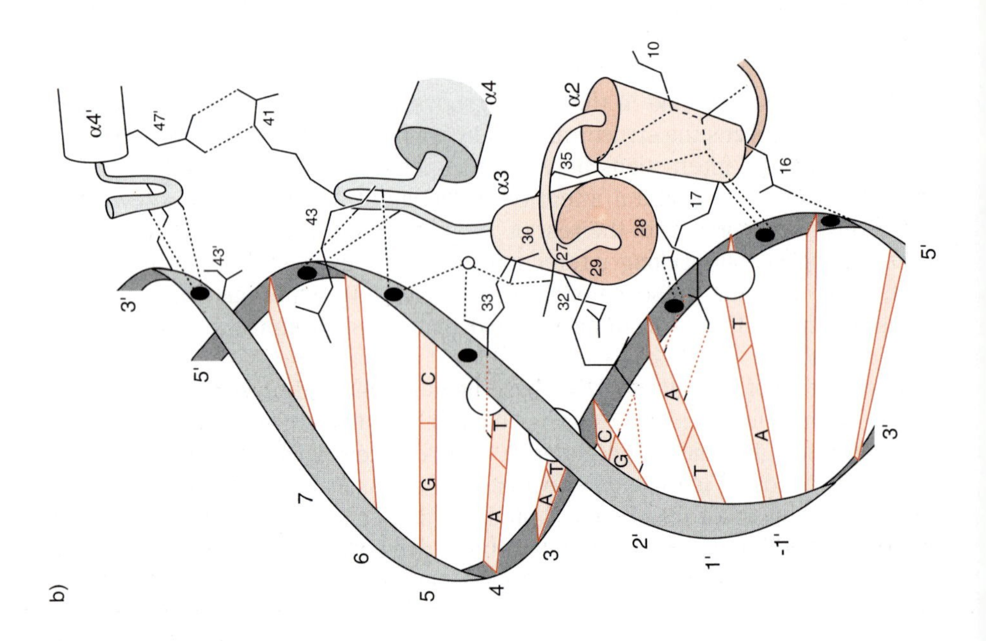
b)
α4'
α4
α3
α2
3'
5'
5'
3'
47'
41
43
43'
33
32
30
29
28
27
35
17
16
10
7
6
5
4
3
2'
1'
-1'

from the protein to recognize specific DNA sequences. In zinc fingers, however, the helix is an integral part of a small structural unit containing the Zn^{2+} ion co-ordinated by two cysteine and two histidine residues.

Zinc fingers are found in many proteins and are specified by a characteristic sequence motif:

$$\text{X-(Tyr/Phe)-X-Cys-X}_{2\text{–}4}\text{-Cys-X}_3\text{-Phe-X}_5\text{-Leu-X}_2\text{-His-X}_{3\text{–}4}\text{-His-X}_4$$

where X is an unspecified amino acid and other hydrophobic residues can occur in place of the tyrosine, phenylalanine, and leucine residues (32). Individually, each finger consists of an α-helix and two β-strands; the Zn^{2+} ion is co-ordinated between the helix and one of the β-strands, and the whole structure is further stabilized by a small core of non-polar residues (*Figure 4.5*). Zinc fingers are normally present in proteins in multiple tandem copies that are linked by a few amino acid residues. Each finger makes contact with a triplet of three adjacent nucleotides of the DNA helix, and adjacent fingers interact with adjacent triplets in the major groove in at least one example (33). Individual fingers can recognize different triplets, so different sequences of nucleotides can be recognized by varying the combinations of zinc fingers present in the protein.

An alternative DNA-binding motif that utilizes zinc occurs in the glucocorticoid and oestrogen receptors. This consists of two α-helices that cross each other at right angles at about their centres, packed together by the interactions of hydrophobic side chains (*Figure 4.6*)(34). The structure can be thought of as a duplication of the loop–helix motif with two Zn^{2+} ions bound. One of the helices functions as the recognition helix, and the other serves primarily a structural role. These proteins are monomeric in solution, but they bind to DNA as homodimers, so binding is co-operative. Their dimerization interfaces are flexible in the monomers, but become structurally complementary upon interacting with the DNA (35). The dimer binds to palindromic DNA sequences.

Figure 4.4. Helix–turn–helix repressor proteins and their interaction with DNA. (**a**) The dimeric structure of the bacteriophage lambda repressor. The helix–turn–helix motifs are shown in orange. The α_3 helix is the recognition helix, which packs against the α_2 helix to maintain its structure. From a drawing kindly provided by B.W.Matthews. (**b**) The interaction of the homologous bacteriophage 434 repressor with DNA. Only part of one monomer of the repressor protein is shown, with the helices depicted as cylinders. The DNA double helix is shown schematically, with the backbone of the individual chains depicted as ribbons and the positions of the phosphate groups by the solid circles. The base pairs are indicated as the flat segments connecting the ribbons; the white spheres on thymidine nucleotides are the methyl groups. Specific interactions between the amino acids of the α_3 recognition helix and the nucleotides exposed in the major groove are shown as orange dotted lines. Other protein–protein and protein–DNA interactions are shown as black dotted lines. Solvent molecules are required as bridges in some of these interactions as shown for the interaction of the side chain of residue 33 with a backbone phosphate group and with the side chain of residue 30. Reproduced with permission from Aggarval, A.K., Rodgers, D.W., Drottar, M., Ptashne, M., and Harrison, S.C. (1988) *Science*, **242**, 899.

A third class of zinc-containing DNA-binding motif is found in proteins of the GAL4 family (36). Here two zinc atoms are co-ordinated by six cysteine residues, with two of the cysteine residues interacting with both zinc atoms. The structure is symmetrical, with each half consisting of an α-helix containing two of the cysteine residues, followed by an extended strand with a third cysteine residue (*Figure 4.7*).

The GAL4 protein is a homodimer and recognizes palindromic DNA sequences. The dimers are held together by a coiled-coil segment of α-helices like those found in the so-called leucine zippers (Section 7.3).

Although the structures of the three types of zinc-containing DNA-binding motifs are not very similar, they all have the common feature that their structures expose an α-helix that fits into the major groove of the DNA to interact directly with the bases of the target sequence.

7.3 Other DNA-binding motifs

Many DNA-binding proteins utilize *h-t-h* or zinc-containing motifs, but not all (27). Members of another class have a DNA-recognition domain at their N

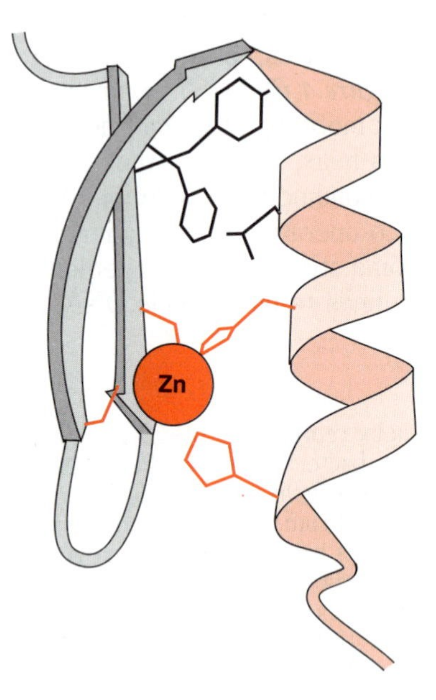

Figure 4.5. Schematic representation of a zinc finger, found in the transcription factor TFIIIA, which consists of a two-stranded β-sheet packed against a recognition helix. The zinc atom is liganded by two cysteine and two histidine side chains, which are shown in orange. Side chains of the non-polar core are also shown. From a drawing kindly provided by D.Neuhaus.

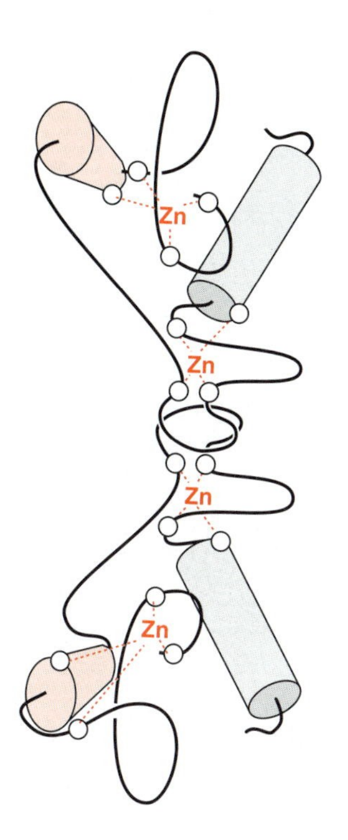

Figure 4.6. The dimeric DNA-binding motif found in the glucorticoid and oestrogen receptors. Each recognition helix is shown in orange. Reproduced with permission from Kaptein, R. (1991) *Curr. Opinion Str. Biol.*, **1**, 63.

terminus linked to a segment of 30 residues that dimerizes to form a two-chain coiled-coil of α-helices, also known as a 'leucine zipper' (see Chapter 1, Section 6.3). The dimer and coiled-coil conformation is formed only above a certain concentration of protein. The coiled-coil serves to link two DNA-recognition domains, which often must be different polypeptide chains, with different DNA-recognition units. The recognition unit is relatively unstructured in the absence of DNA, but becomes α-helical upon binding to the DNA; its detailed structure is not yet known (27).

One other type of interaction, of which there are a few examples, involves recognition of specific DNA sequences by means of β-strands inserted into the major groove of the DNA. An anti-parallel β-ribbon is formed by one β-strand from each of two monomers that dimerize. The β-ribbon protrudes from a primarily helical structural domain and interacts with the major groove of DNA (37).

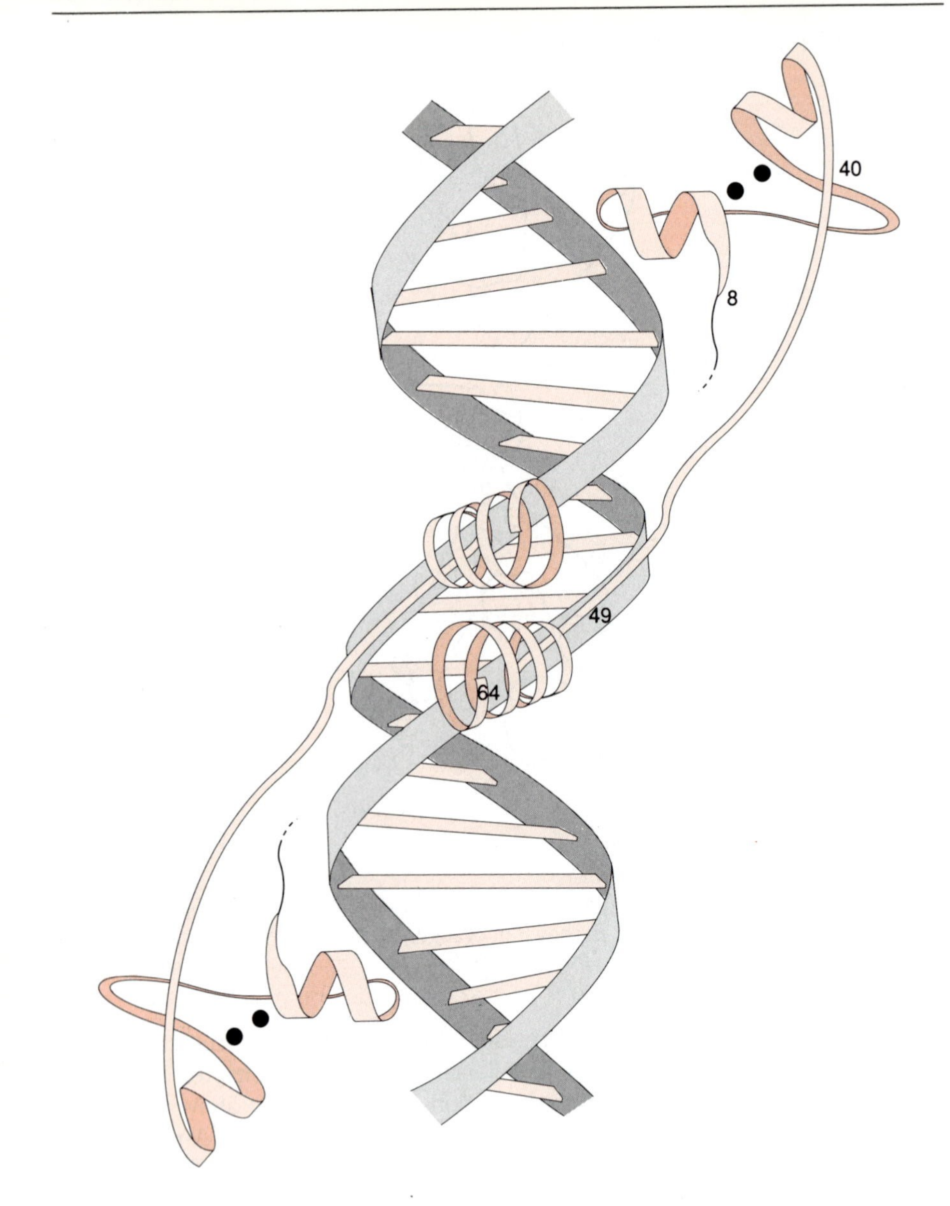

Figure 4.7. Structure of the GAL4–DNA complex. Residues 8–40 comprise the α-helical DNA-recognition element, which interacts with the major groove of the DNA; the two zinc atoms complexed with each recognition element are shown as spheres. Residues 49–64 form the coiled-coil dimerization element, which is linked to the DNA-recognition element by a length of peptide chain that adopts an extended conformation. The linker follows the DNA backbone, its basic residues interacting with the backbone phosphate groups. From a drawing kindly provided by S.Harrison.

8. Regulation of ligand binding

The biological functions of proteins generally require that any binding of ligands is controlled. Most biologically relevant control involves either allosteric regulation or reversible covalent modification of the protein.

8.1 Allosteric regulation

Allosteric regulation allows the affinity of a protein for a ligand at one site to be modified by the binding of a ligand to another site on the protein molecule. *Homotropic* interactions occur when the two ligand molecules are identical, but bind either to different sites on the same polypeptide or to identical sites on different subunits. *Heterotropic* interactions occur when the ligand molecules are different and bind to different sites. Binding of a ligand at one site can increase the affinity of another site, giving *positive co-operativity*, or decrease the affinity, giving *negative co-operativity*. Whenever two or more ligand molecules bind to the same protein molecule, their individual binding interactions are linked functions (*Figure 4.8*), and their effects must be reciprocal. Whatever effect the binding of one ligand has on the affinity of the protein for the other ligand, binding of the other ligand must have the same effect on the protein's affinity for the first.

Allosteric effects in proteins are discussed in terms of two extreme models, generally designated the *sequential* and the *concerted* models (*Figure 4.9*). In the sequential model (38), binding of a ligand at one site on a protein molecule induces a conformational change that is transmitted through the molecule directly to the other ligand-binding sites. This affects their affinities for their respective ligands. Such direct effects can either increase or decrease the affinities of the other sites. There are few limitations on what effects are possible, other than that they must be reciprocal and obey the appropriate linkage relationships (*Figure 4.8*).

The concerted model (39) was based on the observation that allosteric proteins are generally oligomers of identical subunits (although two or more different types of subunits may also be present) and that ligand binding causes a change in quaternary structure in such systems. In this model, two alternative quaternary structures that differ in their affinities for the various ligands co-exist in equilibrium at all times; the transition between them is concerted, that is, it is an all-or-none effect. The quaternary structure that predominates in the absence of ligands usually has low affinity for the primary ligand of the protein, and is designated the *tense* or T state. The alternative structure is designated as the *relaxed* or R state and has high affinity for the primary ligand. As a consequence of the differences in affinities, binding of any ligand will pull the conformational equilibrium between R and T states toward the state with the higher affinity for it. Because there are multiple subunits and binding sites for each ligand, the population of vacant sites with high affinity for that ligand will increase, producing positive homotropic co-operativity. Negative homotropic co-operativity is not possible with the concerted model. However, the R and T states can have reversed affinities for other ligands, so heterotropic allosteric effects can be either positive or negative.

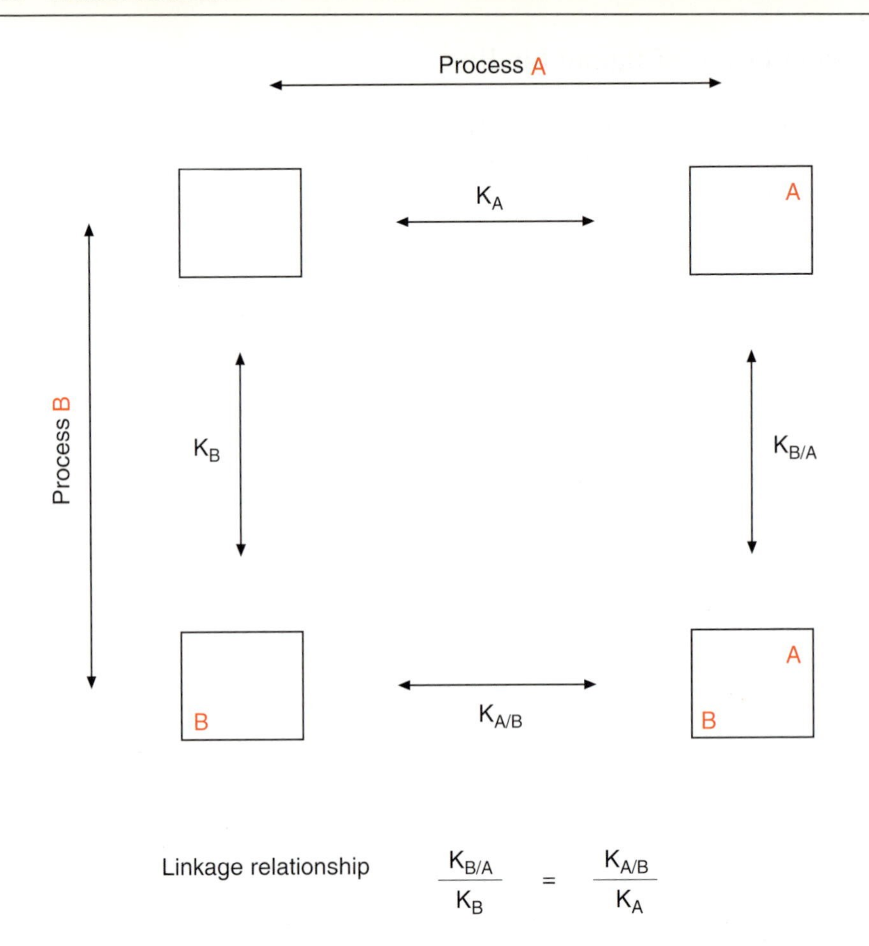

Figure 4.8. General illustration of linkage relationships in a macromolecule, represented as a square. The A and B processes illustrated could, for example, be conformational changes, ligand binding, ionization, disulphide bond formation, etc. The linkage relationship arises because the net free energy change around a closed thermodynamic cycle must be zero. Therefore, if process A affects process B in any way, process B must have the same effect on process A.

The essential difference between the two models of allostery is the way in which the effects of ligand binding at one site are transmitted through the structure to another site. In the sequential model, the effects are transmitted directly from one binding site on one subunit to sites on other subunits. In the concerted model, the interaction occurs through quaternary structure changes, but this requires that the effects of ligand binding are transmitted to the interface between the subunits, and reciprocally, so that ligand binding and the quaternary structure change are coupled.

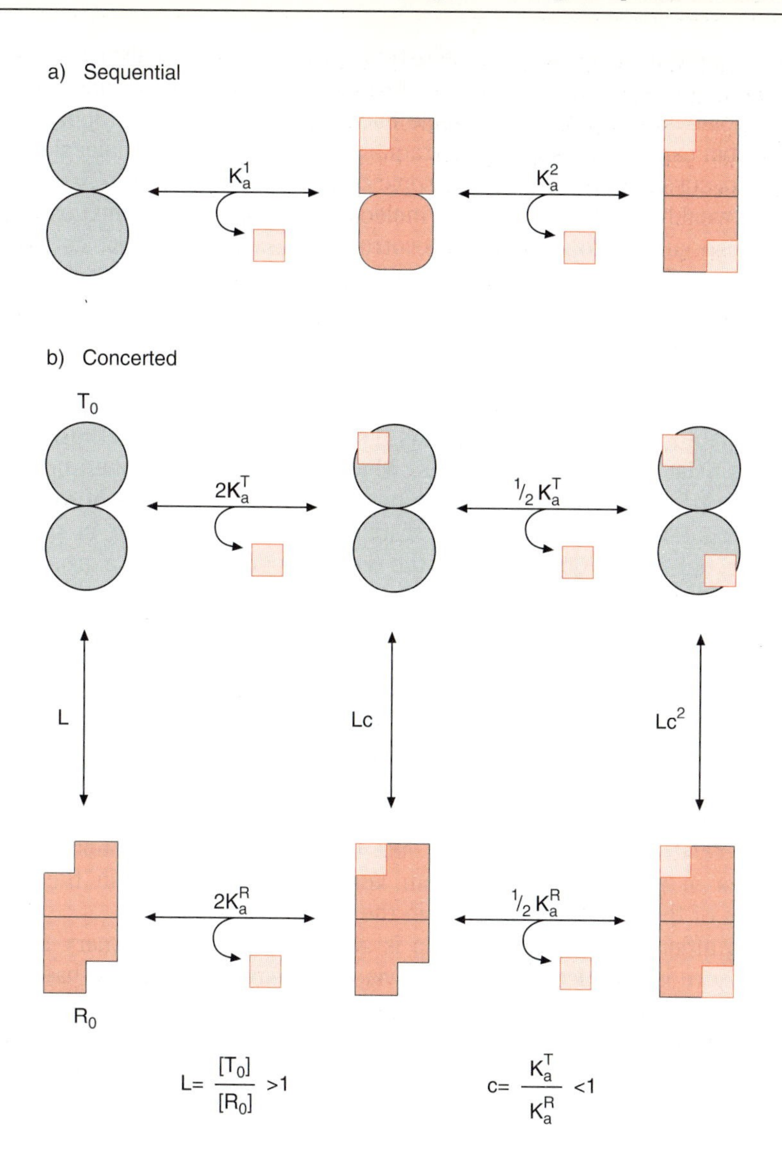

Figure 4.9. The principles of the sequential (**a**) and concerted (**b**) models of allosteric regulation illustrated schematically in a simple dimeric protein. (**a**) In the sequential model, binding of a ligand molecule (small solid square) to one subunit alters the conformation of both subunits to alter the affinity of the other for ligand. (**b**) In the concerted model there are only two alternative conformational states, T and R, in which the intrinsic ligand-binding affinities of individual binding sites (K_a^T and K_a^R) are not affected by binding of ligand to the other subunits. The observed affinities for first and second ligand molecules also depend upon statistical factors resulting from the number of free and bound ligand-binding sites present. Each ligand molecule bound shifts the equilibrium between R and T by a factor c, which is the ratio of the binding constants, K_a^T/K_a^R.

The archetypal example of an allosteric protein is mammalian haemoglobin (40). The α and β subunits of the $\alpha_2\beta_2$ tetramer are homologous, and each binds one O_2 molecule reversibly at a single haem group, so to a first approximation haemoglobin can be considered a tetramer. Binding of four O_2 molecules by the tetramer occurs with positive homotropic co-operativity, since the first O_2 binds relatively weakly and the subsequent molecules have increased affinities. Binding of O_2 is also subject to negative heterotropic interactions, in that its affinity is decreased by the presence of organic phosphates, such as diphosphoglycerate and inositol hexaphosphate, by CO_2, by anions such as Cl^-, and by H^+ ions.

The structural basis of these allosteric effects is relatively straightforward, although many of the details are still not clear (40,41). Haemoglobin exists in two quaternary structures, with the *R* and *T* states corresponding, respectively, to the oxygenated and deoxygenated forms. The tertiary structures of the monomers are very similar in the *R* and *T* states, which differ primarily in their quaternary structures, particularly in the relative positions of the two pairs of αβ dimers. The two quaternary structures are distinct because they differ in respect of which adjacent turn of one α-helix of the α subunit dovetails into a groove of the opposite β subunit; consequently, intermediate quaternary structures are not possible.

The *T* state has the lower affinity for O_2, but higher affinity for the heterotropic ligand effectors. It is the more stable state of haemoglobin without bound O_2. Its affinity for O_2 is low because the small structural changes that take place in the haem group upon binding O_2 are hindered by the tertiary structure of the protein. The more stable the *T* state, the stronger these constraints and the lower the O_2 affinity. These constraints are absent in the *R* state, and the *R* state has relatively high O_2 affinity, like that of an isolated monomer. These differences in O_2 affinity result from very small differences in the tertiary structures of the individual subunits in the *R* and *T* states.

The O_2 affinity of the haem group is sensitive to the quaternary structure because there is a structural coupling, via an α-helix, between the haem group and the interface between subunits that is altered in the *T* to *R* transition. As O_2 is bound, this coupling causes the quaternary structure equilibrium to be shifted towards the *R* state, which predominates in fully oxygenated haemoglobin.

The heterotropic effectors decrease the O_2 affinity simply because they bind more tightly to the *T* state and consequently pull the quaternary equilibrium toward that form, which has low O_2 affinity. For example, organic phosphates bind in the central cavity of the tetramer, at a single site straddling the symmetry axis that is present only in the *T* state; the cavity is too small in the *R* state.

Haemoglobin appears to follow primarily the concerted allosteric model, although it is more complicated because the α and β subunits are not precisely equivalent. Effects that appear to be closer in nature to the sequential model are also evident (40), so elements of both models appear to be used.

Similar mechanisms that involve an equilibrium between two different quaternary structures appear to operate in other allosteric proteins and enzymes, although the details are not as well understood as in the case of haemoglobin. Not all allosteric regulation operates via quaternary structure changes, however, for there are instances of negative co-operativity that require direct interactions between binding sites. In extreme cases, only half the presumably identical sites of an oligomeric protein will bind ligand, but the molecular basis is not known in any instance (42).

8.2 Reversible covalent modification

Binding interactions can be modulated by the covalent modification of proteins. Phosphorylation is the most common type of such modification, but others include adenylylation and ribosylation. All of these examples are similar in that the modification is generally catalyzed by one enzyme and its reversal by another, although in a few instances both reactions are catalyzed by the same enzyme but using different active sites. The mechanisms by which phosphorylation controls ligand binding are now emerging and are likely to prove of general significance in understanding how regulation by covalent modification is achieved.

8.2.1 Regulation by phosphorylation

Phosphorylation at serine, threonine, and tyrosine residues using the γ-phosphate of ATP is the most common reversible covalent modification of proteins and is involved in the regulation of many diverse cellular processes. It commonly occurs in response to an external stimulus, such as binding of a hormone at the cell surface, so phosphorylation complements allosteric regulation, which is primarily sensitive to alterations in intracellular conditions. The external stimulus activates a protein kinase, for example, via changes in the levels of cyclic AMP, to phosphorylate the target protein. Protein phosphatases reverse the modification by hydrolysis to produce inorganic phosphate. The kinases and phosphatases are themselves often subject to complex allosteric regulation and may also be regulated by phosphorylation, to achieve accurate and rapid regulation of the target protein and to avoid futile cycles of ATP hydrolysis.

Two distinct mechanisms of regulation by phosphorylation have been established from X-ray crystallographic studies on phosphorylated and dephosphorylated forms of isocitrate dehydrogenase (IDH) and glycogen phosphorylase (GP). In both instances it is the negative charge of the introduced phosphate group that is of greatest structural significance.

8.2.2 Isocitrate dehydrogenase

IDH is an enzyme of the citric acid cycle that converts isocitric acid to α-ketoglutarate, reducing $NADP^+$ in the process. The activity of IDH is controlled almost solely by phosphorylation and dephosphorylation of the enzyme at Ser-113, which, unusually, are performed by a single enzyme. Phosphorylation does not result in any major conformational change in the protein structure, but there are minor structural rearrangements in the vicinity of the phosphorylated residue. The reason that phosphorylation inactivates the enzyme is that it disrupts the binding of isocitrate at the active site, which normally involves an interaction

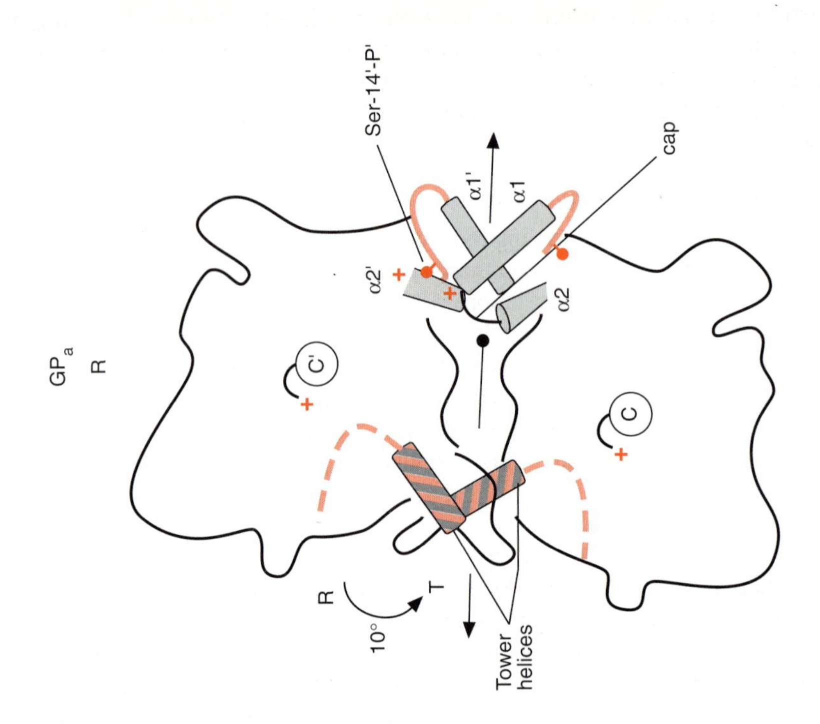
GP$_a$
R
Ser-14'-P'
cap
α1'
α1
α2'
α2
C'
C
10°
R
T
Tower helices

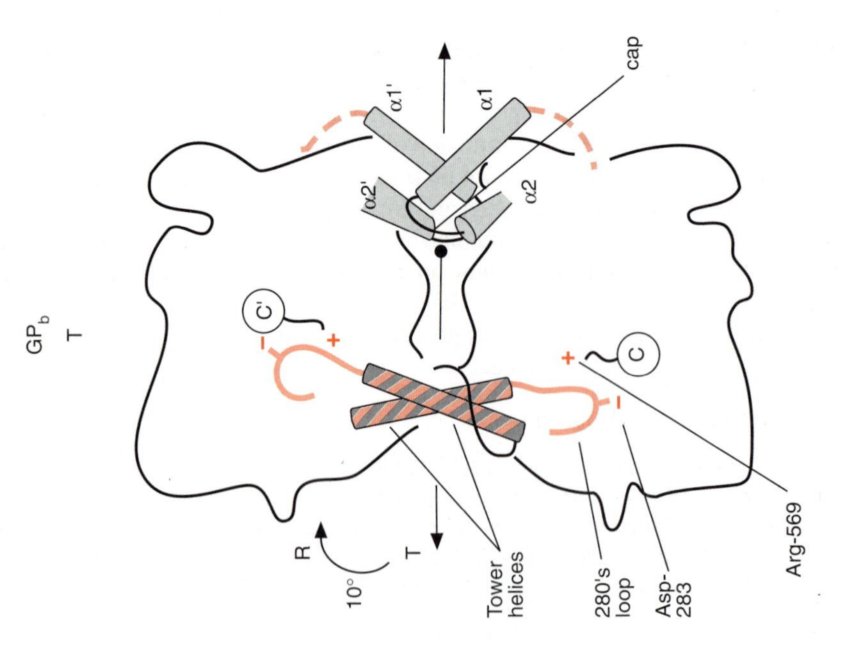
GP$_b$
T
cap
α1'
α1
α2'
α2
C'
C
10°
R
T
Tower helices
280's loop
Asp-283
Arg-569

between the γ-carboxyl of the substrate and the hydroxyl of Ser-113 (43). It is not simply the loss of this interaction in the phosphorylated state that inactivates the enzyme, but also the introduction of a negative charge into the active site which prevents isocitrate entering it by electrostatic repulsion. The effect of phosphorylation can be mimicked by replacing Ser-113 with an aspartic acid or glutamic acid residue, which results in a completely inactive enzyme. However, if this serine residue is replaced by alanine, for example, the enzyme is still partially active.

8.2.3 Glycogen phosphorylase

In contrast to the direct regulation of IDH by phosphorylation at the active site, phosphorylation of GP regulates the enzyme by alterations in its three-dimensional structure (44). The protein is a dimer that occurs in two forms, *a* (GP_a), phosphorylated on Ser-14, and *b* (GP_b), which is not phosphorylated. GP_b is activated by AMP and inhibited by ATP, ADP, glucose, and glucose-6-phosphate and so is regulated primarily by the cell's energy status. This regulation can be overcome by phosphorylation of GP_b by a protein kinase that is activated by elevated cAMP levels in response to neuronal or hormonal stimuli.

The regulation of GP by allosteric effectors and by phosphorylation can be largely explained by the occurrence of two quaternary structure states that correspond to the *R* and *T* states of the concerted allosteric model (Section 8.1). The active *R* state is stabilized by the positive effector AMP, which binds to it preferentially, and by phosphorylation of Ser-14. The phosphorylation induces a conformational change that increases the affinity for AMP and decreases the affinity for the inhibitors.

Crystal structures of the *R* and *T* states of GP indicate how the regulation works. The transition between the *T* and *R* states involves the rotation of one of the subunits by 10° relative to the other about the two-fold axis of symmetry (*Figure 4.10*). At the interface between the subunits, two α-helices, called the 'tower-helices', slide past each other by two turns and alter their angle of tilt to adopt an alternative packing arrangement (*Figure 4.11*). As in other proteins that undergo concerted allosteric transitions, such as haemoglobin (Section 8.1), intermediate quaternary conformations are not possible. The catalytic site lies 15 Å

Figure 4.10. The allosteric transition of GP on activation from the *T* state of GP_b to the *R* state of GP_a. The view is normal to the two-fold axis of symmetry and shows the principal features of the subunit/subunit interface. In the *T* state (left) the N-terminal residues are mobile, and are shown as the orange dashed lines from the α1 and α1′ helices. Conversion to the *R* state (right) by phosphorylation results in the N-terminal residues becoming ordered. Their interaction with the remainder of the protein is promoted by the Ser-14 phosphate group (orange sphere) and tightens the cap/α2′ subunit/subunit interface. One subunit rotates about 10° with respect to the other about the arrowed axis. The change in quaternary structure is accompanied by a change in the tower helices, which pull apart and change their angle of tilt. In the *T* state the 280's loop is ordered and its position blocks access to the catalytic site (C) and inserts an acidic group (Asp-283) into it. In the *R* state this loop of chain is more mobile and the aspartic acid residue is displaced by Arg-569, thus creating the phosphate recognition site for the substrate. Figure kindly provided by L.Johnson.

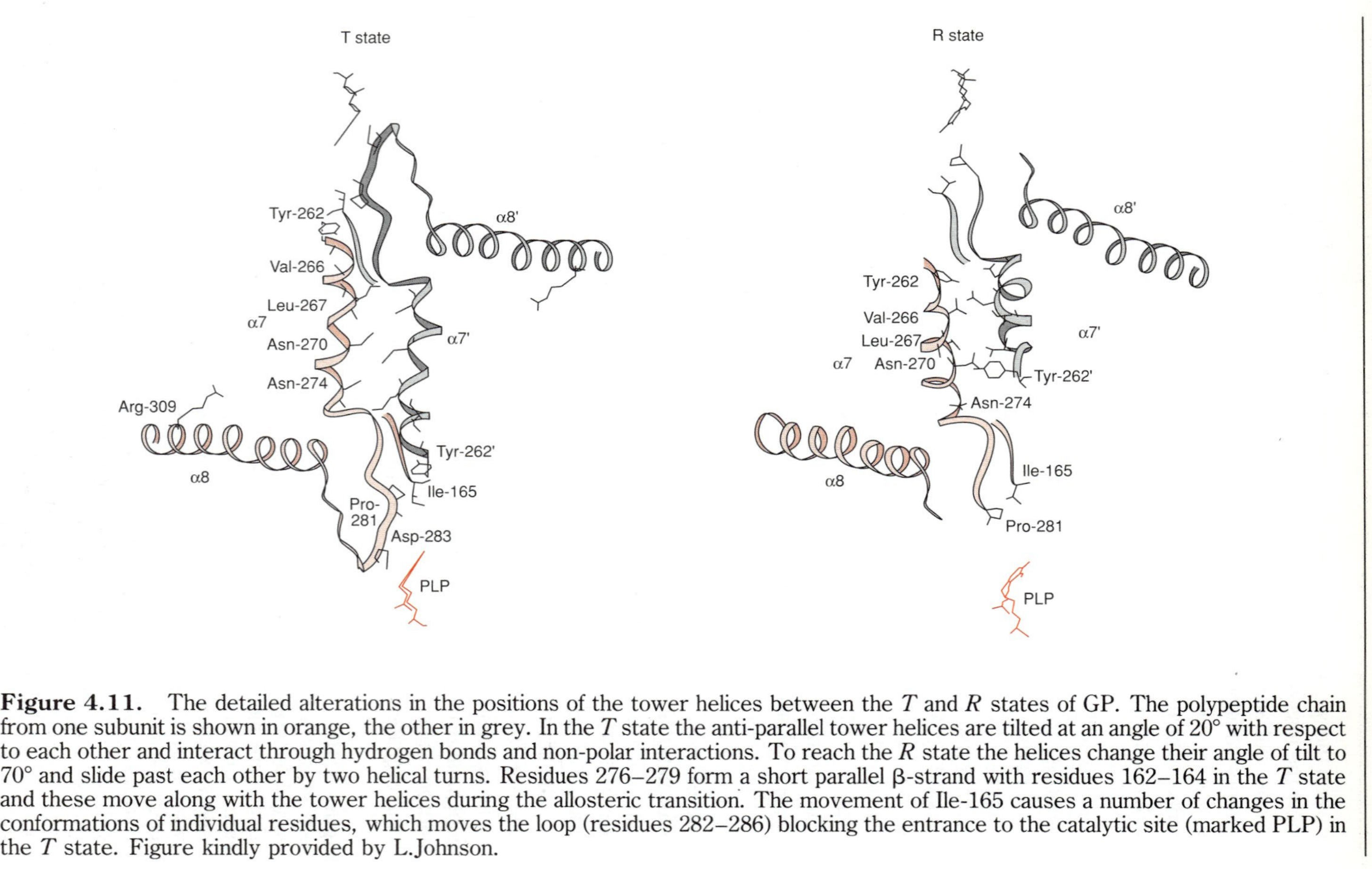

Figure 4.11. The detailed alterations in the positions of the tower helices between the *T* and *R* states of GP. The polypeptide chain from one subunit is shown in orange, the other in grey. In the *T* state the anti-parallel tower helices are tilted at an angle of 20° with respect to each other and interact through hydrogen bonds and non-polar interactions. To reach the *R* state the helices change their angle of tilt to 70° and slide past each other by two helical turns. Residues 276–279 form a short parallel β-strand with residues 162–164 in the *T* state and these move along with the tower helices during the allosteric transition. The movement of Ile-165 causes a number of changes in the conformations of individual residues, which moves the loop (residues 282–286) blocking the entrance to the catalytic site (marked PLP) in the *T* state. Figure kindly provided by L.Johnson.

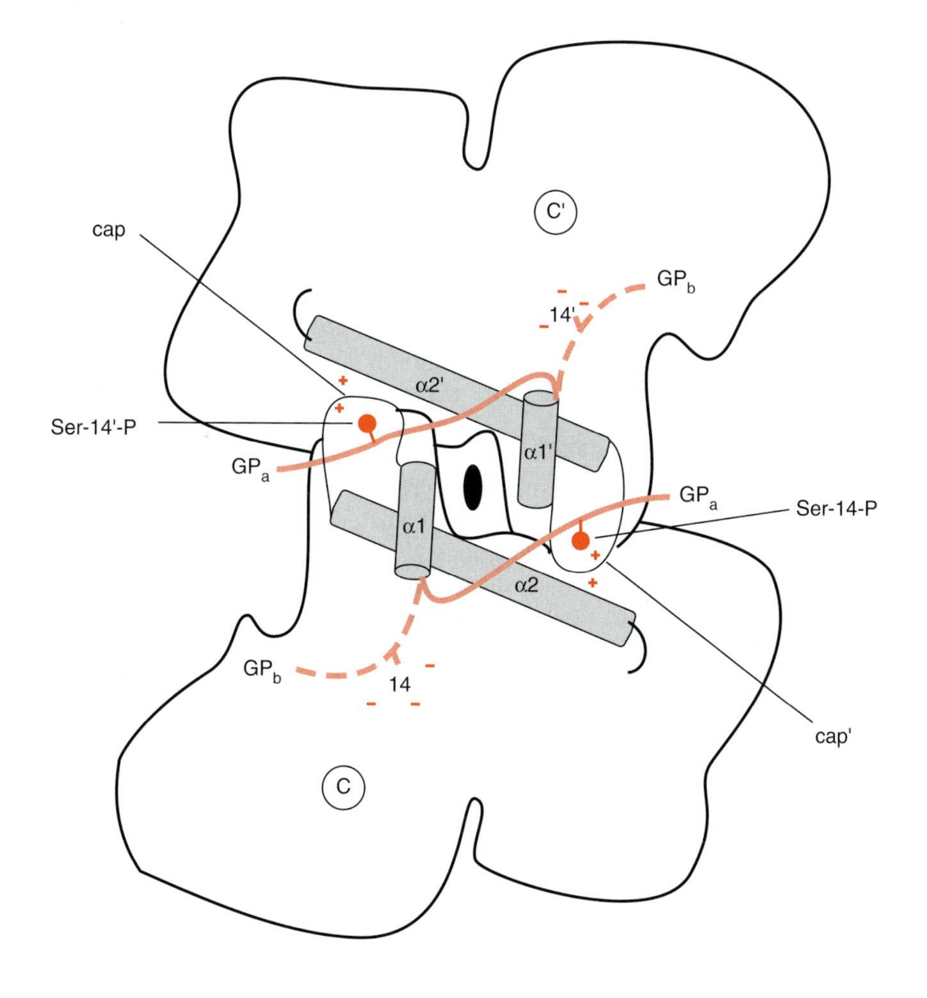

Figure 4.12. Changes in the N-terminal residues on phosphorylation of GP_b to GP_a. The view is down the two-fold axis of the dimer. The N-terminal residues (residues 10–19; shown as the dashed orange line) are mobile in GP_b and located close to negatively charged groups. From residue 19 the chain folds into the α1 helix, followed by the cap and the α2 helix. On phosphorylation the N-terminal residues swing through about 150° and become more ordered (solid orange line). The Ser-14-P group is located at the subunit interface and interacts with two positively charged arginine residues, one from its own subunit, from the α2 helix, and the other from the cap region of the other subunit. These changes are distant from the catalytic site (C). Figure kindly provided by L.Johnson.

from the subunit interface in a cleft, but is connected to the interface by a tower-helix. Movement of the tower-helix during the allosteric transition moves a loop of polypeptide chain that blocks the entrance to the active site in the *T* state, allowing the substrate to enter it in the *R* state. It also causes an arginine residue to displace an aspartic acid residue at the active site, creating a binding site for the phosphate that is necessary in the catalytic process.

Phosphorylation of Ser-14 alters the equilibrium between the *T* and *R* states to favour the latter. With Ser-14 not phosphorylated the N-terminal 16 residues are disordered; after phosphorylation, they adopt a distorted helical conformation (*Figure 4.12*). This probably occurs because the introduction of the negative charge counters the electrostatic repulsions between basic residues at positions 9, 10, 11, and 16 that keep this segment of peptide chain disordered. Adjustments in the tertiary structure of the molecule allow the phosphorylated serine to make favourable electrostatic contacts with arginine residues and some partially exposed non-polar groups to be buried. These changes in the tertiary structure of the protein favour the transition to the *R* state quaternary structure.

8.2.4 Other forms of regulation

The direct and indirect mechanisms of regulation shown by IDH and GP may be generally applicable to other proteins regulated by phosphorylation and to regulation by other covalent modifications. For example, adenylylation of the allosterically regulated protein glutamine synthetase occurs at an interface between subunits (45), which could be consistent with allosteric changes in the quaternary structure being involved. However, other types of structural mechanisms are likely to be involved in the regulation of other proteins, particularly those that are controlled by phosphorylations at multiple sites.

9. Further reading

Berg,J.M. (1990) Zinc finger domains: hypotheses and current knowledge. *Annu. Rev. Biophys. Biophys. Chem.*, **19**, 405.

Boxer,S.G. (1990) Mechanisms of long-distance electron transfer in proteins: lessons from photosynthetic reaction centres. *Annu. Rev. Biophys. Biophys. Chem.*, **19**, 267.

Boyer,P.D. and Krebs,E.G. (1986) Control by phosphorylation. In *The Enzymes*, 3rd edn, Vol. 17. Academic Press, New York.

Crichton,R.R. (1990) Proteins of iron storage and transport. *Adv. Protein Chem.*, **40**, 281.

Heller,A. (1990) Electrical wiring of redox enzymes. *Acc. Chem. Res.*, **23**, 128.

Janin,J. and Chothia,C. (1990) The structure of protein–protein recognition sites. *J. Biol. Chem.*, **265**, 16027.

Steitz,T.A. (1990) Structural studies of protein–nucleic acid interaction: the sources of sequence-specific binding. *Q. Rev. Biophys.*, **23**, 205.

Sweaney,V.W. and Rabinowitz,J.C. (1980) Proteins containing 4Fe-4S clusters: an overview. *Annu. Rev. Biochem.*, **49**, 139.

Williams, R.J.P. (1985) The symbiosis of metal and protein functions. *Eur. J. Biochem.*, **150**, 231.

Wodak,S.J., De Crombrugghe,M. and Janin,J. (1987) Computer studies of interactions between macromolecules. *Prog. Biophys. Mol. Biol.*, **49**, 29.
Wyman,J. and Gill,S.J. (1990) *Binding and Linkage*. University Science, Mill Valley, CA.

10. References

1. Jencks,W.P. (1981) *Proc. Natl. Acad. Sci. USA*, **78**, 4046.
2. Fersht,A.R. (1987) *Trends Biochem. Sci.*, **12**, 301.
3. Chowdhry,V. and Westheimer,R.H. (1979) *Annu. Rev. Biochem.*, **48**, 293.
4. Das,M. and Fox,C.F. (1979) *Annu. Rev. Biophys. Bioeng.*, **8**, 165.
5. Quiocho,F.A., Sack,J.S., and Vyas,N.K. (1987) *Nature*, **329**, 561.
6. Quiocho,F.A., Wilson,D.K. and Vyas,N.K. (1989) *Nature*, **347**, 402.
7. Luecke,H. and Quiocho,F.A. (1990) *Nature*, **347**, 402.
8. Yankeelov,J.A. and Koshland,D.E. (1965) *J. Biol. Chem.*, **240**, 1593.
9. Schulz,G.E., Müller,C.W., and Diederichs,K. (1990) *J. Mol. Biol.*, **213**, 627.
10. Fersht,A.R (1981) *Proc. Roy. Soc. London*, **212**, 351.
11. Chothia,C. *et al.* (1989) *Nature*, **342**, 877.
12. Alzari,P.M. Lascombe,M.-B. and Poljak,R.J. (1988) *Annu. Rev. Immunol.*, **6**, 555.
13. Davies,D.R., Padlan,E.A., and Sheriff,S. (1990) *Annu. Rev. Biochem.*, **59**, 439.
14. Kurosky,A., Barnett,D.R., Lee,T.-H., Touchstone,B., Hay,R.E., Arnott,M.S., Bowman,B.M., and Fitch,W.M. (1980) *Proc. Natl. Acad. Sci. USA*, **77**, 3388.
15. Brandén,C.-I. (1980) *Q. Rev. Biophys.*, **13**, 317.
16. Yamashita,M.M., Wesson,L., Eisenman,G., and Eisenberg,D. (1990) *Proc. Natl. Acad. Sci. USA*, **87**, 5648.
17. Vallee,B.L. and Auld,D.S. (1990) *Biochemistry*, **29**, 5647.
18. Kretsinger,R.H. (1980) *Crit. Rev. Biochem.*, **8**, 119.
19. Strynadka,N.C.J. and James,M.N.G. (1989) *Annu. Rev. Biochem.*, **58**, 951.
20. Churg,A.K. and Warshel,A. (1986) *Biochemistry*, **25**, 1675.
21. Rossmann,M.G., Liljas,A., Brandén,C.-I., and Banaszak,L.J. (1975) In Boyer, P.D. (ed.), *The Enzymes*, Vol. 11. Academic Press, New York, p.61.
22. Bork.P. and Grunwald,C. (1990) *Eur. J. Biochem.*, **191**, 347.
23. Scrutton,N.S., Berry,A., and Perham,R.N. (1990) *Nature*, **343**, 38.
24. Eggink,G., Engel,H., Vriend,G., Terpstra,P., and Witholt,B. (1990) *J. Mol. Biol.*, **212**, 135.
25. Dreusicke,D. and Schulz,G.E. (1986) *FEBS Lett.*, **208**, 301.
26. Schulz,G.E. (1992) *Curr. Opinion Struct. Biol.*, **2**, 61.
27. Harrison,S.C. (1991) *Nature*, **353**, 715.
28. Brennan,R.G. and Matthews,B.W. (1989) *J. Biol. Chem.*, **264**, 1903.
29. Harrison,S.C. and Aggarwal,A.K. (1990) *Annu. Rev. Biochem.*, **59**, 933.
30. Luisi,B.F. and Sigler,P.B. (1990) *Biochim. Biophys. Acta*, **1048**, 113.
31. Brennan,R.G., Roderick,S.L., Takeda,Y., and Matthews,B.W. (1990) *Proc. Natl. Acad. Sci. USA*, **87**, 8165.
32. Klug,A. and Rhodes,D. (1987) *Trends Biochem. Sci.*, **12**, 464.
33. Pavletich,N.K. and Pabo,C.O. (1991) *Science*, **252**, 809.
34. Schwabe,J.W.R. and Rhodes,D. (1991) *Trends Biochem. Sci.*, **16**, 291.
35. Luisi,B.F., Xu,W.X., Otwinowski,Z., Freedman,L.P., Yamamoto,K.R., and Sigler,P.B. (1991) *Nature*, **352**, 497.
36. Marmorstein,R., Carey,M., Ptashne,M., and Harrison,S.C. (1992) *Nature*, **356**, 408.
37. Rafferty,J.B., Somers,W.S., Saint-Girons,I., and Phillips,S.E.V. (1989) *Nature*, **341**, 705.
38. Koshland,D.E., Nemethy,G., and Filmer,D. (1966) *Biochemistry*, **5**, 364.

39. Monod,J., Wyman,J., and Changeux,J.-P. (1965) *J. Mol. Biol.*, **12**, 88.
40. Perutz,M.F. (1989) *Q. Rev. Biophys.*, **22**, 139.
41. Ackers,G.K., Doyle,M.L., Myers,D., and Daugherty,M.A. (1992) *Science*, **255**, 54.
42. Ward,W.H.J. and Fersht,A.R. (1988) *Biochemistry*, **27**, 1041.
43. Hurley,J.H., Dean,A.M., Sohl,J.L., Koshland,D.E., and Stroud,R.M. (1990) *Science*, **249**, 1012.
44. Barford,D., Hu,S.-H., and Johnson,L.N. (1991) *J. Mol. Biol.*, **218**, 233.
45. Yamashita,M.M., Almassy,R.J., Janson,C.A., Cascio,D., and Eisenberg,D. (1989) *J. Biol. Chem.*, **264**, 17681.

Glossary

α-Helix: the regular helical conformation of polypeptide chain in which each residue has approximately $\phi = -57°$, $\psi = -47°$, and there is a hydrogen bond between the NH of each residue and the carbonyl of the residue fourth along in the primary structure.

Accessible surface: the surface defined by the centre of a spherical probe, usually representing a solvent molecule, when it is in van der Waals contact with atoms of the surface of a protein.

Allosteric: resulting from the binding of ligands at two separate sites on a protein molecule; the binding sites can be different or be identical but on different subunits.

Amphipathic: having both a hydrophilic and a hydrophobic part.

Anomalous scattering: resonance between the vibrations excited by incident X-rays and the natural oscillations of the inner electrons of many heavy atoms that causes the scattering of X-rays to be non-symmetrical.

β-Sheet: collection of β-strands lying side-by-side, linked together in either a parallel or anti-parallel manner by hydrogen bonds between their backbone NH and carbonyl groups.

β-Strand: segment of polypeptide chain with an extended conformation that is part of a β-sheet.

Chiral: not superimposable on its mirror image.

Circular dichroism (CD): spectral phenomenon arising from a difference in absorption of left- and right-polarized light by chiral groups of atoms.

Conformations: three-dimensional arrangements in space of atoms of a molecule, differing only in rotations about covalent bonds and, possibly, disulphide bonds.

Contact surface: van der Waals surface of atoms of a molecule that is in contact with molecules of the solvent.

Convergent evolution: evolutionary change that gradually reduces the difference between two genes or proteins that were originally unrelated, as a result of selection for similar structural and functional properties.

Co-operativity: any phenomenon in which the magnitude of one interaction is dependent upon the existence and strengths of one or more other interactions; as a result molecules with incomplete sets of interactions are less stable than expected from the sum of all the interactions when present simultaneously.

Coulomb's law: the electrostatic interaction between two point charges is determined by the product of the two charges, divided by the distance between them and the dielectric constant (or relative permittivity) of the environment.

Divergent evolution: evolutionary change that gradually increases the differences between two genes or proteins that originated from a common ancestor.

Domain: part of a protein with a folded conformation that is relatively independent structurally from the remainder of the protein molecule; when comprised of a continuous segment of polypeptide chain, the isolated segment often can maintain this folded conformation.

Exon: segment of a gene that is transcribed into the precursor mRNA and remains in the mature mRNA after processing and removal of introns.

Halophilic: functioning normally at high salt concentrations.

Heat capacity: a measure of the amount of energy required to raise the temperature of a system; it is proportional to the structural changes that take place upon changing the temperature and to the temperature dependence of the enthalpy and entropy of the system.

Homologous: being similar as a result of a common evolutionary origin.

Homopolypeptide: polypeptide chain obtained by condensation in peptide bonds of the same amino acid, dipeptide, or other short peptide.

Hydrogen bond: attraction between a nucleophilic atom (primarily N or O in proteins) and the hydrogen atom covalently bonded to a second nucleophilic atom; the attraction is primarily electrostatic in nature, but with some covalent nature in very strong interactions.

Hydrophilic: having affinity for water; the magnitude of this effect is usually measured by partitioning from the gas phase to aqueous solution.

Hydrophobic: preferring non-polar environments over an aqueous environment, generally measured by partitioning between the two; the term generally refers to non-polar atoms or groups.

Intron: intervening sequence; linking portion of a gene that is translated, but is spliced out of the initial RNA transcript and is not present in the mature mRNA.

Ligand: any molecule that interacts with one or a few specific sites on a protein.

Linked functions: any two equilibria in a molecule that can occur at the same time; any effect that one has on the other must be reciprocal.

Macrostate: an ensemble of individual microstates, such as specific three-dimensional conformations of a polypeptide chain, that are rapidly interconverted and are not distinguished individually.

Mesophilic: functioning normally at moderate temperatures and conditions.

Mosaic proteins: proteins consisting of multiple structural and functional units that were probably assembled by a genetic process of 'domain shuffling'.

Molten globule: a third conformational state of proteins (in addition to folded and unfolded) that is stable in certain proteins under particular conditions; it is relatively compact and contains significant amounts of secondary structure, but relatively little or no fixed tertiary structure.

Non-polar: having no net charge and no asymmetric charge distribution and consequently being chemically unreactive.

Peptide: a few amino acids linked sequentially by peptide bonds; the maximum number of residues in a peptide is not defined, but the properties of a peptide are usually those expected from the sum of the constituent parts.

Polar: having a non-uniform charge distribution and being chemically reactive.

Polypeptide chain: a string of amino acids linked sequentially via peptide bonds.

Primary structure: the linear sequence of amino acid residues in a polypeptide chain.

Prosthetic group: a ligand that tends to remain bound to a protein and to participate in the protein's biological function.

Quaternary structure: the three-dimensional arrangement of multiple polypeptide chains in a folded protein; it may also be used to refer to the arrangement of individual folded domains.

Ramachandran plot: a two-dimensional plot of all the possible combinations of ϕ and ψ, the two torsion angles of the polypeptide backbone of an amino acid residue, showing the permitted values and those adopted by residues in proteins and in regular conformations.

Random coil: Conformational macrostate of a polymer comprised of many interconverting individual conformations of the entire molecule in which the conformation of each monomer is independent of the conformations of all monomers distant in the covalent structure.

Salt bridge: a close interaction between two groups with opposite net charge; the most stable salt bridges probably also have an element of hydrogen bonding in the interaction.

Secondary structure: local regular conformations of polypeptide chain, especially α-helices, β-strands, and reverse turns.

Tertiary structure: stable overall three-dimensional structure of a folded polypeptide chain.

Thermophilic: functioning normally at high temperatures.

van der Waals radius: the radius of an atom at which the density of its electron cloud is sufficient to prevent a second atom coming closer than the sum of the van der Waals radii of the two atoms.

van der Waals surface: the surface of a molecule defined by the van der Waals radii of the constituent atoms.

Index